Improving the Continued Airworthiness of Civil Aircraft

A Strategy for the FAA's Aircraft Certification Service

Committee on Aircraft Certification Safety Management
Aeronautics and Space Engineering Board
Commission on Engineering and Technical Systems
National Research Council

NATIONAL ACADEMY PRESS
Washington, D.C. 1998

NATIONAL ACADEMY PRESS • 2101 Constitution Avenue, N.W. • Washington, DC 20418

NOTICE: The project that is the subject of this report was approved by the Governing Board of the National Research Council, whose members are drawn from the councils of the National Academy of Sciences, the National Academy of Engineering, and the Institute of Medicine. The members of the panel responsible for the report were chosen for their special competencies and with regard for appropriate balance.

This study was supported by the Federal Aviation Administration under contract No. DTFA03-96-C-00041. Any opinions, findings, conclusions, or recommendations expressed in this publication are those of the author(s) and do not necessarily reflect the view of the organizations or agencies that provided support for the project.

International Standard Book Number: 0-309-06185-7

Available for sale from:

National Academy Press
Box 285
2101 Constitution Ave., N.W.
Washington, DC 20055
800-624-6242
202-334-3313 (in the Washington Metropolitan Area)
http://www.nap.edu

Printed in the United States of America

COMMITTEE ON AIRCRAFT CERTIFICATION SAFETY MANAGEMENT

JAMES G. O'CONNOR (chair), Pratt & Whitney (retired), Coventry, Connecticut
M. CRAIG BEARD, Federal Aviation Administration (retired), Delaplane, Virginia
EUGENE E. COVERT, Massachusetts Institute of Technology, Cambridge
THEODORE E. DUMONT, Sikorsky Aircraft (retired), Milford, Connecticut
FRANK C. FICKEISEN, The Boeing Company (retired), Bellevue, Washington
CLYDE KIZER, Airbus Service Company, Herndon, Virginia
DEAN J. LENNARD, General Electric Aircraft Engines (retired), Cincinnati, Ohio
STEVEN R. LUND, Boeing Commercial Airplane Group, Douglas Products Division, Long Beach, California
C. JULIAN MAY, Delta Airlines (retired), Kennesaw, Georgia
WILLIAM H. SCHULTZ, General Aviation Manufacturers Association, Washington, D.C.
NOZER D. SINGPURWALLA, George Washington University, Washington, D.C.
COLIN TORKINGTON, Air Navigation Commission of the International Civil Aviation Organization, Montreal, Canada

ASEB Liaison

WILLIAM HOOVER, U.S. Air Force (retired), Williamsburg, Virginia

Staff

ALAN ANGLEMAN, Study Director
GEORGE LEVIN, Director, Aeronautics and Space Engineering Board
MARY MESZAROS, Senior Project Assistant
TED MORRISON, Senior Project Assistant

AERONAUTICS AND SPACE ENGINEERING BOARD

Staff

Foreword

The National Research Council (NRC) was asked to conduct an independent assessment of the safety management process used by the Aircraft Certification Service of the Federal Aviation Administration (FAA) to define how the current process might be improved. The Committee on Aircraft Certification Safety Management, comprised of individuals chosen for their diverse perspectives and technical expertise in accordance with established procedures of the NRC, undertook the assigned study. The Committee was comprised of six members with industry expertise in large aircraft manufacturing and operations (engines, airframes, avionics, maintenance, and safety disciplines), two members with related experience (rotary wing and general aviation aircraft), and four members outside industry altogether (International Civil Aviation Organization, the FAA, and academia).

The committee was asked to review common causes of accidents and incidents involving civil aircraft and determine which causes might be related to the certification process with an emphasis on continued airworthiness. The focus of the study was on how an already small frequency of accidents, made smaller still by a necessary connection with the certification process, might be made even smaller in the next decade. The committee had to consider quantifiable, qualitative, and latent risks and, based on risk assessment methodologies used by manufacturers, determine how the small risk of accidents could be further reduced. Defining a top-level aircraft certification safety management process that could reduce near-term accident risks entailed taking into account expected changes in both the aircraft fleet and certification. The committee was also asked to consider how their recommended safety management process might be applied to civil transport aircraft and other types of aircraft, to identify implementation barriers, and to define a strategy for assessing the effects of the recommended approach.

A complex study like this one, which investigates near-term safety improvements in just one of many processes that could affect aircraft safety, requires that committee members be in possession of the relevant facts and prior experience to make informed judgments. The inclusion of such committee members, especially those with industrial experience, was deemed essential for the credibility of the results among manufacturing and operational constituencies, as well as the FAA. In addition, it was also necessary to ensure that the committee maintain a balance with a number of outside members, so that the results would be unbiased. Public credibility on matters of safety depends on this balance. We believe we have struck a careful balance in the composition of this expert committee. Moreover, this report was carefully reviewed and critiqued, according to standard NRC procedures, by independent and knowledgeable experts from diverse perspectives. The responsibility for the final report rests entirely with the authoring committee and the institution.

The consensus recommendations in this report outline improvements to the certification process that could effect possible near-term improvements to a system that is already quite safe. The recommendations call for a deeper partnership between the FAA and the manufacturing industry and operational community. Based on data now in the possession of industry, the partnership would facilitate the analysis of these data under a process for which the FAA would have oversight responsibility.

We commend this report to serious consideration by the FAA and industry. We believe the report makes a major contribution to the enhancement of aviation safety.

WILLIAM WULF, *President*
National Academy of Engineering

Preface

Every day, the United States air transportation system provides safe and efficient service to millions of travelers, and it is the common goal of U.S. airlines, aircraft manufacturers, and the federal government to make air travel even safer. Accomplishing this goal, even as the number of passengers and total miles flown increases each year, will require cooperation among many different organizations.

The Aircraft Certification Service of the Federal Aviation Administration (FAA) oversees aspects of civil aviation safety related to the design and manufacture of aircraft and aircraft systems, equipment, and parts. This includes assessing accidents, incidents, and other unexpected events to determine when certification standards for new and existing aircraft should be modified to maintain expected levels of continued airworthiness. As one element of the overall effort to improve aviation safety, the FAA requested that the National Research Council conduct an independent assessment of the safety management processes used by the Aircraft Certification Service. In response, the National Research Council established the Committee on Aircraft Certification Safety Management to conduct the study, the results of which are published in this report.

As described herein, the committee verified that the current aircraft certification system contributes to the low rate of accidents in this country, as evidenced by the small fraction of accidents caused by aircraft system malfunctions. Even so, the current approach to aircraft certification and continued airworthiness can be improved. A new approach could be established that accounts for—and takes advantage of—changes in the aircraft industry since the current system was established. By becoming more performance-based, the Aircraft Certification Service could leverage existing resources to carry out its stated missions and priorities more effectively and help the FAA meet its extremely challenging safety goals. Industry should fully participate in the safety management process in partnership with the FAA. Safety monitoring and preventive measures should be based on reliable data and modern analysis techniques, tools, and logic. Barriers to implementation of a new, more effective safety management process are mostly bureaucratic and legalistic and will be difficult to overcome in a timely fashion.

This report has been reviewed by individuals chosen for their diverse perspectives and technical expertise, in accordance with procedures approved by the National Research Council's Report Review Committee. The purpose of this independent review is to provide candid and critical comments that will assist the authors and the National Research Council in making the published report as sound as possible and to ensure that the report meets institutional standards for objectivity, evidence, and responsiveness to the study charge. The content of the review comments and draft manuscript remain confidential to protect the integrity of the deliberative process. We wish to thank the following individuals for their participation in the review of this report:

Bruce Aubin, Air Canada (retired)
Robert Blouin, National Business Aircraft Association
Anthony Broderick, Federal Aviation Administration (retired)
Robert Davis, The Boeing Company
John Lauber, Airbus Service Company
Robert G. Loewy, Georgia Institute of Technology
Duncan Luce, University of California-Irvine
Stuart Matthews, Flight Safety Foundation
Kenneth Rosen, Sikorsky Aircraft Corporation
Harvey Schadler, General Electric Corporate Research and Development Center (retired)
Gareth Thomas, University of California-Berkeley

While the individuals listed above have provided many constructive comments and suggestions, responsibility for the final content of this report rests solely with the authoring committee and the National Research Council.

The committee also wishes to thank everyone else who supported this study, especially those who took the time to participate in committee meetings (see Appendix C).

JAMES G. O'CONNOR, Pratt & Whitney (retired)
Chairman, Committee on Aircraft Certification Safety Management

Contents

Tables and Figures

TABLE

FIGURES

Executive Summary

As part of the national effort to improve aviation safety, the Federal Aviation Administration (FAA) chartered the National Research Council to examine and recommend improvements in the aircraft certification process currently used by the FAA, manufacturers, and operators. The Committee on Aircraft Certification Safety Management was formed to execute this task, which included the following key elements:

- define an improved approach for managing risk and promoting the safety of U.S. civil aircraft, with a focus on the continued airworthiness of large transport airplanes
- identify barriers to implementing the recommended approach and how they might be overcome
- discuss the special needs of general aviation and rotorcraft

BACKGROUND

The safety of major U.S. airlines is unmatched by comparable modes of public transportation. The effectiveness of the aircraft certification process is an important factor contributing to this successful record. Major organizations involved in the aircraft certification process are the FAA, manufacturers, and operators. The aircraft certification process encompasses three primary elements:

- rulemaking and policy development: defining and implementing new and modified regulations and associated policy guidelines for use by the FAA and industry
- certification: issuing new and amended type certificates, production certificates, and airworthiness certificates for new and modified aircraft, engines, and other equipment
- continued airworthiness and other activities related to continued operational safety: verifying the ongoing safety of products manufactured in accordance with approved designs by monitoring existing aircraft

The Aircraft Certification Service (AIR)[1] is the department within the FAA that has lead responsibility for carrying out these actions. Specific functions include issuing initial and amended type certificates for designs of new and derivative aircraft, supplemental type certificates (STCs) for designs of modifications to existing aircraft, production certifications to certify a manufacturer's ability to build an aircraft in accordance with an approved design, airworthiness certificates for individual aircraft verifying that they have been manufactured in accordance with approved designs, and airworthiness directives to correct unsafe conditions in existing aircraft.

Federal Aviation Regulations generally do not stipulate how certification standards should be met because design processes typically do not lead to a single "best" solution to meet a given set of certification standards. Writing effective regulations that focus on the characteristics of systems and aircraft instead of on specific design procedures can be difficult unless current technology, and how it is likely to evolve in the future, is well understood.

The most important function of aviation safety management is to prevent accidents. As shown in Figure ES-1, the "primary" causes of most accidents are associated with human error. For example, controlled-flight-into-terrain and loss-of-control-in-flight accidents, which almost by definition involve human factors, account for more than half of all fatal accidents. Aircraft system malfunctions, on the other hand, are involved in a relatively small fraction of aircraft incidents and accidents. However, it is also true that most accidents are caused by a chain of events, any one of which could have prevented the accident. This provides multiple opportunities to improve safety. By addressing individual factors in the chain of events, the accident rate can be reduced.

[1]"AIR" is the designation used by the Federal Aviation Administration for the Aircraft Certification Service.

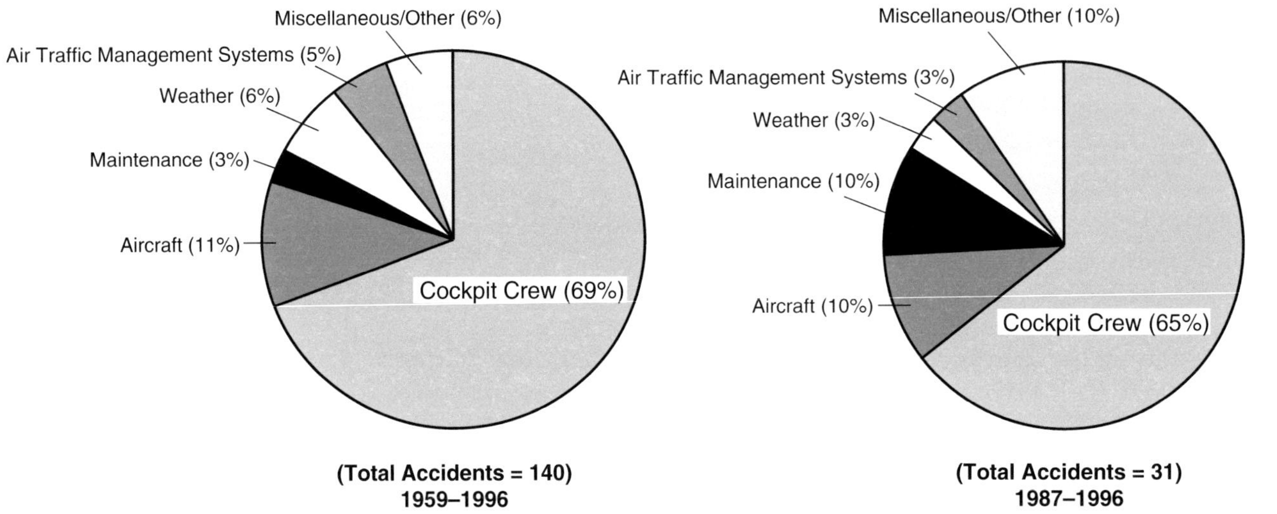

FIGURE ES-1 Primary cause factors for hull loss accidents involving large U.S.-registered commercial jet airplanes. Source: Boeing 1997.

The Major Finding. The recommended safety management process should improve the ability of the FAA/AIR, manufacturers, and operators to take corrective action based on incident data—before an accident takes place—and to set priorities based on assessments of current and future risk. However, the current process is already highly effective—as indicated by the small contribution of aircraft system malfunctions to the overall accident rate—and changes to the current system must be carefully structured to avoid unintended consequences that might reduce safety in some situations.

PRIORITIES

Major Recommendation 1. It is critically important that the FAA and AIR conduct business in a new fashion with regard to aircraft certification and continued airworthiness. As an essential first step, AIR should revise its budget and manpower allocations to better reflect its mission priorities, which are as follows:

1. continued airworthiness and other activities related to continued operational safety
2. rulemaking and policy development
3. certification

The vast majority of aircraft that will operate during the next 10 years have either already been manufactured or will be manufactured to already certificated design specifications. Monitoring the safety of operating aircraft is essential to obtain a true picture of safety, to detect and resolve problems as soon as possible, and to validate airworthiness standards. Improvements in standards for initial type certification are typically based on lessons learned from the continued airworthiness process. Therefore, making the continued airworthiness process more effective is essential to improving safety in the near term and providing the foundation for long-term improvements. The primacy of this task is acknowledged in the FAA's stated priorities. Currently, however, AIR's type certification activities receive more resources than the other two areas combined.

SAFETY MANAGEMENT PROCESS

Major Recommendation 2. It is essential that the FAA improve its safety management process. The FAA should work with the operators and manufacturers of large transport airplanes and engines to define and implement a proactive process that includes the following elements and tasks:

Key Elements
- data collection
- database management
- risk analysis
- risk management/action
- monitoring effectiveness

Specific Tasks
- Manufacturers, with the advice and consent of operators and the FAA, should define data requirements and processes for sharing data. Comprehensive flight operations quality assurance systems similar to the British Airways Safety Information System (BASIS) should be used as a starting point.
- Operators should provide required data, as agreed upon.
- Manufacturers should solicit data from additional sources, such as the National Transportation Safety Board, International Civil Aviation Organization, and National Aeronautics and Space Administration, to augment the operational database.

- Manufacturers, with oversight from the FAA and the assistance of operators, as required, should collect, organize, and analyze data to identify potential safety problems.
- Manufacturers should recommend corrective action for potential safety problems and seek consensus by operators. The FAA should make sure that actions proposed by manufacturers and operators will be effective, making regulatory changes and mandating compliance, as appropriate.
- Manufacturers and operators, with oversight from the FAA, should monitor the effectiveness and timeliness of corrective action and the safety management process (see Figure ES-2).

The thrust of this recommendation is that industry should collect, organize, and analyze safety data and take appropriate corrective action to protect the safety of the fleet. The FAA should not independently collect, organize, or analyze safety data for large transport aircraft. Instead, the FAA should oversee the entire process, providing direction, assessing the accuracy and objectivity of industry's risk analyses, and mandating corrective action, as appropriate. The overall objective is to produce a more effective safety management process that routinely monitors operations and maintenance, uses data on incidents and other abnormalities to identify potential hazards proactively, and takes corrective action before hazards cause an accident.

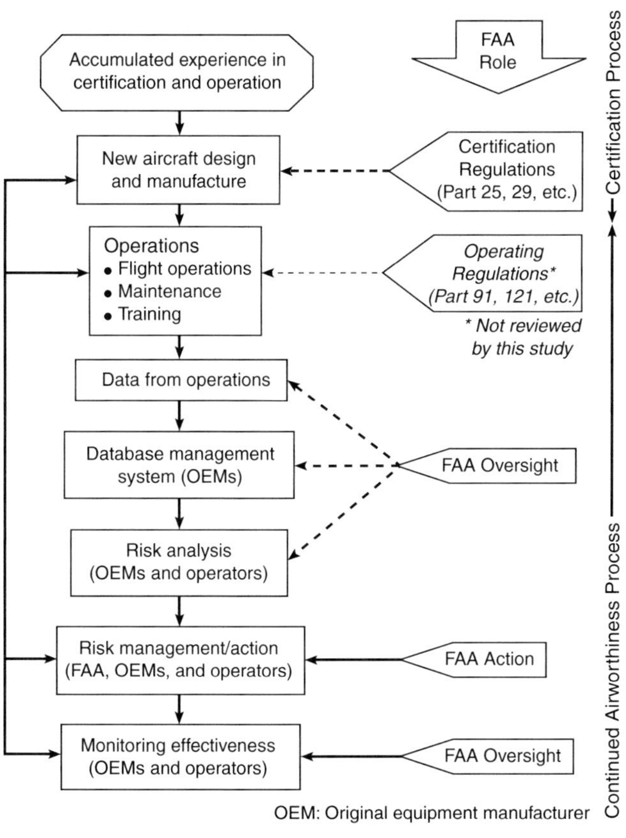

FIGURE ES-2 The recommended process for aircraft certification safety management.

Many systems are currently used by industry and the FAA for generating, collecting, and storing data. Many of these data systems are not coordinated or used effectively, however, consuming scarce resources that could and should be applied to other safety-related activities. Some of the databases cannot be fully utilized because of poor data quality and difficulties in interpretation. Most existing data collection and monitoring systems have not been formulated to identify hazards that may arise from unusual combinations of factors that may not individually present a significant hazard. To establish a more proactive safety management system, a database management process is needed that focuses on accurately identifying precursors to potential accidents. Such a process would rely heavily on incident data. Currently, the FAA does not have access to detailed information about many incidents throughout the world.

The current safety management process lacks a widely accepted risk analysis system or methodology. Such a methodology is necessary to establish credible priorities for effective and timely resource allocation and action. With regard to risk management/action, the basic elements are already in place but must be enhanced to attain desired improvements in aviation safety. Currently, there is a tremendous backlog of pending regulatory actions, including hundreds of airworthiness directives issued by foreign airworthiness authorities. There is no legal requirement for U.S. operators of the affected aircraft to implement any of these directives unless the FAA concurs that the action should be mandatory and issues an equivalent airworthiness directive. The size of the backlog could be reduced if more FAA personnel were dedicated to reviewing foreign directives. In this regard, the FAA is not following its own stated safety priorities.

A method for accurately assessing the effectiveness of the safety management process is important because remedial action can disrupt airline operations and reduce the competitive standing of operators and manufacturers. Accurate information on the effectiveness of remedial action would put the FAA in a better position to justify its own priorities and allocate resources to areas with the highest potential for improving aviation safety. A proactive safety management process would frequently recommend action in response to incidents, before an accident occurs. Demonstrating the effectiveness of safety actions will be important for building confidence in future recommendations.

Significant improvements in the current safety management process would be greatly facilitated by better cooperation among federal agencies, operators, and manufacturers. The committee believes that the overriding role of the FAA should be to provide encouragement and leadership in the United States and internationally to maximize industry participation, to implement a more standardized global system, and to overcome the barriers that will hinder implementation of the recommended process. In parallel with efforts to make appropriate regulatory changes, the FAA should expeditiously negotiate binding letters of agreement with manufacturers and operators to implement as much of the recommended safety management process as possible.

In developing the recommended safety management process, the committee considered several possible approaches for improving aviation safety. For example, the current safety management process, which has achieved an excellent safety record, could be continued with only minor changes. Another possibility would be for a single organization to collect and analyze safety data for all types of aircraft instead of sharing this responsibility among many different organizations. The committee also considered how much of the process should be voluntary and how much should be mandated. The committee believes the recommended safety management process draws appropriately from these options and provides practical guidance for enhancing the safety of U.S. civil aviation.

APPROVED DESIGN ORGANIZATIONS

Major Recommendation 3. AIR should promote aircraft safety by certifying the competency of applicants' design organizations rather than relying on the FAA's ability to detect design deficiencies through spot checks. The FAA should work with industry and Congress to obtain legislative and regulatory authority in a timely fashion to do the following:

- Certificate and rate approved design organizations (ADOs) and invest them with the responsibility for ensuring that applications for type certificates, type certificate amendments, STCs, technical standard order authorizations (TSOAs), and parts manufacturer approvals (PMAs) comply with applicable airworthiness standards. ADOs would be required to have the technical capabilities necessary for competently approving designs only within the limitations of their rating.
- Require ADOs and holders of production certificates to collect and analyze relevant safety data received from operators and to define corrective action in the event unsafe conditions are detected.
- Require applicants for design approvals to either hold an ADO certificate or employ the services of an ADO.
- As an interim step, give higher priority to the ongoing rulemaking action that would increase organizational delegation to manufacturers of large aircraft and engines under the FAA's current legislative authority. The FAA already uses this authority to grant organizational delegation to manufacturers of small aircraft and engines.

Existing legislation and regulations do not require applicants for type certificates, STCs, and other design approvals (TSOAs and PMAs) to show that they have the technical qualifications to develop a safe design or to conduct the engineering evaluations and certification tests necessary to show compliance with applicable airworthiness standards. There are no requirements for type certificate or STC applicants to establish or maintain a technical organization to monitor, evaluate, and propose corrective action in response to operator reports of safety problems for which they are responsible. The current process unrealistically assumes that spot checks by the FAA during reviews of new designs and design changes will reveal all items of noncompliance with airworthiness standards. The current process also limits the ability of the FAA to take advantage of the capabilities of certificated design organizations. The present system requires the FAA to spend considerable resources on "false starts" by applicants, particularly STC applicants, that do not have the technical qualifications to complete the engineering process required for design approval.

The committee believes that safety would be enhanced if the FAA focused its design approval process on determining that applicants' design organizations are technically qualified and have internal review processes that ensure compliance with the applicable airworthiness standards, rather than continuing to rely on its own ability to determine compliance through spot checks of the applicant's analyses and tests. The FAA should examine the technical qualifications and integrity of design organizations, including their understanding of regulations and policies and their ability to properly implement them. Qualified organizations should then be certificated as ADOs, allowing them to make detailed findings of compliance in accordance with published policies. FAA audits would verify continued compliance, in part by ensuring that ADOs' level of involvement in specific projects is appropriate in light of the technical issues involved.

Establishing a system of ADOs would reduce FAA resources required to conduct its certification functions, making additional resources available for continued airworthiness activities as recommended by Major Recommendation 1.

The FAA is working with industry to develop regulatory changes that would delegate additional certification functions to industry. The committee urges the FAA to continue working in this direction as an interim step toward the certification of applicants' design organizations. However, a more comprehensive restructuring of the process is needed to implement the ADO concept envisioned by the committee. This restructuring would require legislative authorization in the form of changes to Title 49 of the U.S. Code.

HUMAN FACTORS

Major Recommendation 4. The FAA should support and accelerate efforts (1) to define the minimum data required by the flight crew to maintain adequate situational awareness during all phases of flight and reasonable emergency scenarios and (2) to determine how this data can be presented most effectively.

Human factors issues, specifically human errors, are significant contributors to most incidents and accidents. Improving the situational awareness of flight crews and air traffic controllers and improving the effectiveness of maintenance personnel are essential for preventing most serious incidents and accidents associated with human error. Thus, it is important that the FAA harness the increasing body of human factors knowledge being developed by other organizations. That is one of the tasks of the FAA's Human Factors Study Group. The FAA should ensure that this group includes strong representation in the fields of cognitive science and basic neuroscience so that it can form a cohesive framework for understanding the very large number of human factors studies that are now being conducted, especially with regard to cockpit design. Training is another important tool for reducing many types of human error. However, training was outside the scope of this study and, thus, it is not addressed in the report.

BARRIERS

Major Recommendation 5. In order for AIR to contribute as much as possible to improvements in aviation safety, the FAA—in partnership with industry, Congress, the Department of Transportation, and other involved parties—must take aggressive action to overcome barriers associated with the following:

- external pressures and influences faced by the FAA
- coordination and communications within the FAA
- legal issues
- the rulemaking process
- the economic impact of proposed changes to the safety management process

There are a number of barriers, both internal and external, that will make it difficult to implement the recommended safety management process. The FAA must overcome these barriers to achieve current national goals for improving aviation safety over the next 10 years.

External pressures and influences faced by the FAA. Highly publicized accidents are often caused by factors not associated with the greatest aviation hazards. Political and public pressure to "solve" these highly publicized accidents can divert attention, personnel, and funds from efforts to address more significant risk factors. Crisis management and the impetus to take quick action can also result in action that is less effective in the long run than taking the time to develop a more effective initial response. As a first step toward reducing the negative impact of external pressure on the safety management process, the FAA should work with other responsible agencies to educate the public better about ongoing efforts to improve aviation safety. However, the committee believes that fully addressing this issue is likely to require major organizational changes, such as establishing a senior interagency communications or safety management board, that were beyond the scope of this study.

Coordination and communications within the FAA. Many of the organizational elements in the FAA enjoy considerable autonomy over their assigned areas of responsibility and lack an effective means of communicating and resolving differences. More effective communications—within AIR and between AIR and other FAA offices, such as the Flight Standards Service—would considerably improve the aircraft certification safety management process by facilitating the exchange of information within the FAA and the dissemination of complete and consistent information to industry.

Legal issues. Legal issues are associated with the potential for public disclosure of sensitive information under the Freedom of Information Act, the possibility of regulatory enforcement against individuals or companies who voluntarily disclose information about safety problems, and increased exposure to legal liability arising from the litigation discovery process in an environment where more data is collected, stored, and shared. The FAA has significant leeway in addressing the first two problems by modifying internal policies and the Federal Aviation Regulations. Legal discovery issues are likely to remain a significant problem unless Congress enacts legislation to protect voluntarily shared data from the threat of discovery action directed toward parties with whom such data are shared.

Rulemaking process. The FAA's rulemaking process is defined by internal policies and, more importantly, by legislation and executive branch regulations the FAA cannot waive unilaterally. As currently implemented, the rulemaking process quite often takes 5 to 10 years. Although timely action is possible, especially in the case of highly publicized or emergency safety actions, many worthwhile, safety-related activities linger without action for unreasonably long periods of time. These delays are a significant safety issue. During 1998, the FAA plans to implement recommendations from an internal study on how to improve the efficiency of the FAA rulemaking process. This is a positive first step, but much more needs to be done. The Department of Transportation, other executive branch agencies, and Congress should also work with the FAA to modify legislation, directives, and regulations, as necessary, to substantially improve the responsiveness of the rulemaking process.

Economic impact of proposed changes to the safety management process. The air transport industry is highly competitive, and this natural competitiveness is a potential barrier to the voluntary sharing of data required to implement the recommended safety management process. Manufacturers and operators bear the cost of making safety improvements, and their support will be forthcoming only to the extent that the identified risks are credible and the corrective action seems reasonable in terms of effectiveness and cost. The FAA should work with industry to develop confidence in the cost/benefit analyses used to justify changes in the safety management process. The FAA should also subsidize pilot projects by operators and manufacturers to validate the cost effectiveness of new systems for data collection, database management, and analysis.

SMALL AIRPLANES AND ROTORCRAFT

Major Recommendation 6. Plans to implement the recommended safety management process within the small airplane and rotorcraft communities should be developed in cooperation with small airplane and rotorcraft operators, manufacturers, and associations of operators and manufacturers. The FAA should establish cooperative agreements that define the roles of individual operators, individual manufacturers, their associations, and AIR. These agreements should define the following:

- responsibilities of operators for submitting data
- responsibilities of operators, manufacturers, associations of operators and manufacturers, and AIR for data collection, database management, risk analysis, risk management/action, and monitoring effectiveness
- processes for the routine exchange of data and risk analysis results between operators, manufacturers, associations, and AIR to facilitate effective risk management/action
- a publicity program to inform the small airplane and rotorcraft communities of the new safety management process

There are many differences between large transport airplanes and small airplanes. The number of small aircraft registered in the United States and the number of pilots licensed to fly them far exceeds the numbers of large transport airplanes and pilots. On average, large transport airplanes and pilots are in the air for many more hours per year than small airplanes and pilots. The pilots of small airplanes have a much wider range of experience and skills and operate in a much broader spectrum of functional modes than the pilots of transport airplanes. Small airplanes and rotorcraft are used for recreation, sightseeing, pipeline patrols, scientific experimentation, crop dusting, and firefighting.

Thousands of civil rotorcraft are registered in the United States, most of them operating as general aviation aircraft. Rotorcraft operating characteristics and some of the missions they undertake, such as logging, law enforcement, and emergency rescue, create a risk environment that is quite different from the risk environment of most fixed wing airplanes. Rotorcraft and small airplanes also operate out of many more airports and landing areas than large airplanes, and many of these airports do not have control towers or other landing and takeoff aids.

The safety management process for small airplanes and rotorcraft must be flexible enough to accommodate the diverse nature of these communities, and this is likely to be a difficult challenge. Final accident investigation reports for small airplanes and rotorcraft show that the majority of accidents are attributable to human error, and the small role played by aircraft system malfunctions indicates that the current aircraft certification and continued airworthiness process is working well.

REFERENCES

Boeing. 1997. Primary Cause Factors for Hull Loss Accidents Involving Large U.S. Registered Commercial Jet Airplanes. Seattle, Wash.: Boeing Commercial Airplane Group.

1

Introduction

The air transportation system in the United States is safer than comparable modes of public transportation. For major airlines (i.e., air carriers operating under Part 121 of the Federal Aviation Regulations [FARs]), the average number of fatalities per 100 million passenger miles is about 0.7, compared to about 1.8 for automobiles, about 11 for intercity buses, and about 17 for trains (BTS 1998a, 1998b, 1998c, 1998d; NHTSA, 1996). In terms of safety, travel on major airlines within the U.S. is matched only by travel on major airlines in other highly developed countries. Nevertheless, fatal accidents are always tragic, and complacency on the part of the FAA or industry is not an appropriate response. In fact, the FAA has already established a strategic goal of zero accidents.

The Federal Aviation Administration (FAA) plays a major role in promoting aviation safety.[1] However, the FAA will face several important challenges in the future. If the aircraft accident rate remains constant or slowly decreases, the annual number of accidents will swell as the number of flights increases to meet consumer demands. The public has the right to expect high levels of safety, and it is incumbent upon industry and the FAA to improve the effectiveness of their safety programs. In part, this means reacting to major accidents by taking aggressive action to prevent similar accidents, but without detracting from ongoing safety programs to address other risks.

Almost all aircraft accidents are caused by a chain of events, the elimination of any one of which could have prevented the accident. The most common link in these chains involves human factors (pilots, air traffic controllers, maintenance crews, etc.). However, in some cases, one or more links in the accident chain are associated with the design of the aircraft. Either a design deficiency results in an equipment malfunction that leads to an accident, or a design enhancement could have prevented an unexpected event from resulting in an accident.

The FAA's Aircraft Certification Service (AIR)[2] is responsible for promoting the safety of new aircraft by certifying that they meet established safety standards. Certification includes type certificates (certification of all-new aircraft designs), amended type certificates (certification of derivative aircraft designs based on previously certificated products), production certificates (certification of a manufacturer's ability to produce aircraft in conformance with a certificated design), and airworthiness certificates (certification of the airworthiness of each newly manufactured aircraft). AIR also promotes the continued airworthiness of existing aircraft by mandating modifications when operating experience indicates the presence of a real or potential hazard.[3]

As part of the FAA's efforts to improve aviation safety, AIR chartered the National Research Council to examine safety-related elements of the certification and continued airworthiness process and to recommend an approach to improve AIR's risk evaluation and risk management. In response, the National Research Council's Aeronautics and Space Engineering Board formed the Committee on Aircraft Certification Safety Management. This report is the result of the study conducted by that committee. A complete list of the committee's findings and recommendations appears in Appendix A.

OBJECTIVES

The study statement of task required that the National Research Council conduct an independent study that would accomplish the following goals:

[1]Throughout this report, the word "promote" is often used to describe the FAA's role with regard to safety. As directed by legislation and Supreme Court rulings, the FAA *promotes* safety by overseeing industry activities. Every aircraft manufacturer, operator, repair facility, etc., is individually responsible for *ensuring* safety by providing products and services that are safe and comply with regulatory standards (see Chapter 2).

[2]"AIR" is the designator used by the Federal Aviation Administration for the Aircraft Certification Service.

[3]A more complete description of AIR's roles and priorities appears in Chapters 2 and 4.

1. Develop an understanding of the most common causes of accidents and incidents encountered by civil aircraft (i.e., large transports, small airplanes, and rotorcraft) and determine whether these causes are related to the aircraft certification process, with special emphasis on the continued airworthiness of aircraft. Base the inquiry on available data from industry, investigative organizations, and regulatory agencies; and include factors such as manufacturing standards and airworthiness directives. Consider accidents and incidents in the last 10 years, at a minimum. More importantly, consider how accidents can be prevented in the next 10 years.
2. Develop an understanding of the ability of the current aircraft certification process to identify three kinds of risks: well quantified risks, qualitatively understood risks, and latent risks.
3. Develop an understanding of the risk assessment methodologies used by a representative set of manufacturers of aircraft and aircraft systems.
4. Define the key elements of a top-level aircraft certification safety management process that could reduce the risk of accidents in the next 10 years. Take into consideration expected changes in the commercial aircraft fleet, as well as the operational and economic effects of changes to the U.S. certification process on the aviation manufacturing industry, aircraft owners and operators, flight crews, and regulatory agencies in the United States and abroad.
5. Identify the elements of the recommended safety management process that are applicable to civil transport airplanes and describe how the process should be modified for other types of aircraft (i.e., small airplanes and rotorcraft).
6. Define potential barriers to implementing the recommended safety management process and how they might be overcome.
7. Define a strategy for assessing the effects of the recommended safety management process.

The vast majority of aircraft that will be operated in the next 10 years either have already been manufactured or will be manufactured to already certificated designs. Changes in the standards for certification are typically based on lessons learned from the continued airworthiness process. Therefore, the committee focused on continued airworthiness issues (i.e., the airworthiness directive system and AIR's role in establishing new or amended rules for aircraft design certification). The scope of this study did not include other safety-related topics, such as the process used by the FAA's Flight Standards Service to monitor compliance with airworthiness directives; the role of individual offices within the FAA; administrative procedures; training of flight crews; ground-based air traffic management systems; flight operating procedures; or the certification and monitoring of pilots, air carriers, maintenance facilities, etc. (which are the responsibility of the Flight Standards Service).

STUDY PROCESS AND APPROACH

The members of the Committee on Aircraft Certification Safety Management had expertise in aircraft design, manufacture, operations, maintenance, and certification; aviation safety; accident investigation; and risk management. Biographical sketches of committee members appear in Appendix B.

To accomplish its task, the full committee met five times for discussions with personnel from regulatory agencies, manufacturers, operators, and pilots and for private deliberations. Small groups of committee members conducted additional fact-finding trips to meet with representatives of the rotorcraft industry, the European Joint Aviation Authorities (JAA), and various organizations in Washington, D.C. Participants in these meetings are listed in Appendix C. The committee collected, reviewed, and discussed a great deal of information provided by the FAA and industry, including information on the FAA's Safety Performance Analysis System, National Aviation Safety Management Program, and the Aviation Safety Initiative Review.

To fulfill its charge, the committee had to come to grips with the interactions among flight safety, certification, and the application of new technology. These three elements can be unsnarled by understanding the role of the regulatory authority. Regulations generally do not stipulate how certification standards should be met. In part, this is because design processes typically do not lead to a single "best" solution to meet a given set of certification standards. However, writing effective regulations that focus on the characteristics of systems and aircraft instead of specific design procedures can be difficult unless the state of the art is well understood. Maintaining a high level of understanding is a constant challenge for engineers in many disciplines; it is especially challenging for personnel, such as FAA regulators, who generally are not directly involved in research and technology development.

As the air transportation system evolves and public use of the system increases, the perception of risk by the public, industry, and government also evolves. In preparing this report, the committee has attempted to define a practical approach that will facilitate an orderly evolution of standards and procedures and will be consistent with public expectations and the technical capabilities of both the FAA and the aircraft industry.

This report often refers to the three major participants in the aircraft certification and continued airworthiness process: manufacturers, operators, and the FAA. Each of these participants includes many smaller constituencies, each with its own special concerns. Improving the safety management process will, in many cases, require changes in regulations, corporate policies, and labor contracts. A cooperative approach to formulating these changes will ensure that they are broadly endorsed and can be implemented in a timely fashion. This will require balancing the concerns of all constituencies, including FAA managers and inspectors;

manufacturers of aircraft, engines, and systems; airline managers, pilots, and mechanics; managers of independent repair facilities; members of the traveling public, and others.

ORGANIZATION OF THIS REPORT

This report is focused on the committee's primary task: defining the key elements of an improved aircraft certification safety management process for large transport airplanes and recommending how potential barriers to implementing the recommended process could be overcome. Chapter 2 provides background information on the role of AIR, on how that role has evolved, and on specific regulatory actions that are part of the current safety management process. Chapter 3 provides background information on the causes of incidents and accidents.

Chapter 4 describes the recommended safety management process, which includes a mechanism for monitoring its own effectiveness. Chapter 4 focuses on improving the continued airworthiness of large transport airplanes, and it also includes several recommendations for improving the effectiveness of the certification process. Chapter 5 describes the relationships between human factors, environmental factors, and aircraft systems in accidents and incidents, followed by comments on several current initiatives to reduce accidents and incidents associated with human error. Chapter 6 recommends approaches for overcoming five key barriers to the implementation of the recommended safety management plan. Chapter 7 describes special characteristics of the small aircraft and rotorcraft communities and the special concerns that these characteristics raise with respect to the recommended safety management plan.

The appendices contain supplemental information. Appendix A is a summary list of all findings and recommendations. Appendix B contains short biographies of all committee members. Appendix C contains a list of participants in committee meetings. Appendix D contains additional information on probability and the application of reliability analysis tools, such as fault tree analysis, to aviation. Appendix E is an example of a "knowledge base" system (i.e., a knowledge-based database system). Appendix F is a sample legislative amendment that would authorize the certification of approved design organizations.

REFERENCES

BTS (Bureau of Transportation Statistics). 1998a. Airlines: Scheduled Service Accidents, Fatalities and Rates. U.S. Department of Transportation. Online. Bureau of Transportation Statistics Homepage. Available: http://www.bts.gov/NTL/data/tables/chap09/tab9-5/tab9-5.txt. June 6, 1998.

BTS. 1998b. Average Length of Haul, Domestic Freight and Passenger Modes. U.S. Department of Transportation. Online. Bureau of Transportation Statistics Homepage. Available: http://www.bts.gov/programs/btsprod/nts/chp1/tbl1x12.html. June 6, 1998.

BTS. 1998c. Bus Profile. U.S. Department of Transportation. Online. Bureau of Transportation Statistics Homepage. Available: http://www.bts.gov/programs/btsprod/nts/apxa/bus98.html. June 6, 1998.

BTS. 1998d. Railroad Passenger Safety Data. U.S. Department of Transportation. Online. Bureau of Transportation Statistics Homepage. Available: http://www.bts.gov/programs/btsprod/nts/chp3/tbl3x35.html. June 6, 1998.

NHTSA (National Highway Traffic Safety Administration). 1996. Motor Vehicle Occupants and Nonoccupants Killed and Injured, 1986–1996. U.S. Department of Transportation. Online. Available: http://www.nhtsa.dot.gov/people/ncsa/overvu96.html#Summary. June 6, 1998.

2

Role of the Aircraft Certification Service (AIR)

AIR (the Aircraft Certification Service) is the department within the FAA that develops and administers safety standards for aircraft and related products that are manufactured in the United States or are used by operators of aircraft registered in the United States. Related products include engines, propellers, equipment, and replacement parts. This chapter provides background information that is necessary to understand the recommended safety management process and other findings and recommendations that appear in Chapters 4 through 7 of this report. Readers who are already knowledgeable about AIR's roles and processes may wish to skip this chapter.

AIR's functions include issuing initial and amended type certificates for designs for new aircraft, issuing supplemental type certificates (STCs) for designs of modifications to existing aircraft, issuing production certificates to certify a manufacturer's ability to build an aircraft in accordance with an approved design, and issuing airworthiness certificates to verify that individual aircraft have been manufactured in accordance with approved designs and are in safe operating condition. AIR is also responsible for overseeing the continued operational safety of manufacturers and aircraft. AIR issues airworthiness directives (ADs) to correct unsafe conditions when they are detected in aircraft that have previously been issued an airworthiness certificate. AIR is not responsible for oversight of pilots, airlines, or other facets of aviation operations that are the purview of the FAA's Flight Standards Service. Additional information on AIR's activities and how they apply to engines, spare parts, etc. appears in the section on Specific Regulatory Actions near the end of this chapter.

AIR's mission priorities are as follows (FAA, 1998):

1. continued airworthiness and other activities related to continued operational safety
2. rulemaking and policy development
3. certification

Continued airworthiness activities are the highest stated priority because they have the greatest immediate impact on the safety of operating aircraft and because they promote the continued satisfactory performance of approved systems, such as manufacturers' approved quality control systems. Rulemaking and policy development are considered to be a higher priority than issuing new certificates because the integrity of the certification program depends on the currency of applicable rules and policies.

HISTORY AND STATUTORY AUTHORITY

No person may lawfully operate a civil aircraft in the United States unless it has an airworthiness certificate. The requirement for certificating civil aircraft dates back to enactment of the Air Commerce Act of 1926 (Komons, 1978), which is the origin of the FAA's current process for aircraft certification and continued airworthiness. The Air Commerce Act of 1926 also established an Aeronautics Branch within the Department of Commerce. The Aeronautics Branch subsequently evolved into the FAA, which is organizationally within the Department of Transportation.

During the early years of U.S. aviation, the government did not play a formal role in promoting the safety of civil aircraft. The Aeronautics Branch issued the first civil aviation safety regulations, the "Air Commerce Regulations," which became effective on December 31, 1926. The Air Commerce Regulations included the first standards for licensing (certification) of aircraft. On March 29, 1927, the Aeronautics Branch first certificated a civil aircraft: the Buhl-Verville Model J-4 Airster, an open cockpit, single engine, bi-wing aircraft also known as the model C-3A. The Air Commerce Regulations thus became the cornerstone of the U.S. airworthiness standards. Those standards, and the processes used to implement them, have evolved over the past 72 years through legislative and regulatory changes, reorganizations among executive branch departments, and reorganizations of individual agencies. This evolution has been driven by the following factors:

- continued high levels of public and congressional concerns about air transportation safety
- the introduction of new technologies, which have advanced the efficiency of the air transportation system and provided opportunities to improve aviation safety
- lessons learned from investigations of civil aviation accidents and incidents
- changes in international air transportation regulations and policies

The Air Commerce Act of 1926 was superseded by the Civil Aeronautics Act of 1938, which in turn was superseded by the Federal Aviation Act of 1958. In July 1994, the Act of 1958 was recodified as Subtitle VII of Title 49 of the United States Code, which currently provides the FAA's regulatory authority. Airworthiness standards, known as the FARs (Federal Aviation Regulations), are separately codified in Chapter I of Title 14 of the Code of Federal Regulations.[1] Among other things, FARs set forth the type certification requirements—known as "airworthiness standards"—for aircraft designs, the requirements for manufacturers' production quality control systems, the requirements for airworthiness certification of individual aircraft, and the operations and maintenance rules for air carriers and repair facilities. Both AIR and the Flight Standards Service administer these regulations. The parts of the Code of Federal Regulations (i.e., the FARs) that are most relevant to the roles of AIR and the Flight Standards Service are listed below.

- Definitions and General Procedures
  - Part 1. Definitions and abbreviations
  - Part 11. General rulemaking procedures
  - Part 13. Investigative and enforcement procedures
- FARs Administered by AIR
  - Part 21. Certification procedures for products and parts
  - Part 23. Airworthiness standards for normal, utility, acrobatic, and commuter category aircraft. Note: Part 23 includes certification requirements for all types of small airplanes. A design for a multi-engine, propeller-driven airplane with 19 or fewer seats (excluding seats for the flight crew) and a maximum certificated takeoff weight of 19,000 pounds or less must be type certificated as a "commuter category airplane" under FAR 23 unless it is certificated under Part 25.
  - Part 25. Airworthiness standards for transport category airplanes. Note: Unless eligible for certification under Part 23, all airplane designs with 10 or more seats (excluding seats for the flight crew) or a maximum certificated takeoff weight of more than 12,500 pounds must be type certificated under Part 25. However, any multi-engine airplane design, regardless of size, may be type certificated as a transport category airplane if it meets the airworthiness standards of FAR 25.
  - Part 27. Airworthiness standards for normal category rotorcraft
  - Part 29. Airworthiness standards for transport category rotorcraft
  - Part 33. Airworthiness standards for aircraft engines
  - Part 35. Airworthiness standards for propellers
  - Part 39. Airworthiness directives
- FARs Administered by the Flight Standards Service
  - Part 43. Maintenance, preventive maintenance, rebuilding, and alteration
  - Part 91. General operating and flight rules. Note: These rules are applicable to all aircraft, but general aviation aircraft are often referred to as Part 91 aircraft because they are not subject to additional operating rules under Parts 119, 121, 125, or 135. General aviation aircraft may be of any type and size, as long as they are not operated in accordance with Part 121 or scheduled operations under Part 135. General aviation operations are mostly recreational flying, but also include business aviation, flight instruction, and industrial flying, such as firefighting, aerial application of pesticides, etc.
  - Part 119. Certification requirements for air carriers and commercial operators. Note: Part 119 defines when Parts 121, 125, and 135 are applicable.
  - Part 121. Operating requirements for domestic, flag, and supplemental operations. Note: Part 121 generally applies to airlines operating jet airplanes or any other types of airplanes with 10 or more passenger seats or a payload capacity of 7,500 pounds or more. Aircraft used for operations covered by Part 121 usually must be certificated under Part 25 or as commuter category aircraft under Part 23.
  - Part 125. Certification and operations for aircraft with a seating capacity of 20 or more passengers or a maximum payload capacity of 6,000 pounds or more. Note: Part 125 is primarily applicable only when common carriage is not involved, and it is not applicable to aircraft operated under Parts 121 or 135.
  - Part 135. Operating requirements for commuter and on-demand operations. Note: Part 135 is generally applicable to rotorcraft and fixed-wing aircraft with nine or fewer passenger seats and a payload capacity of 7,500 pounds or less. Because of practical and regulatory considerations, most Part 135 aircraft are certificated under Part 23. Most Part 135 operators of large aircraft have been required to transition to Part 121 operations.
  - Part 145. Repair stations

[1]Title 49 of the United States Code is defined by federal legislation and can only be changed by new legislation. Title 14 of the Code of Federal Regulations is defined by regulations issued by federal agencies, such as the FAA, in accordance with relevant sections of the United States Code.

As indicated in this list, AIR is responsible for FARs associated with certification, airworthiness standards, and ADs; and the Flight Standards Service is responsible for FARs associated with operations and maintenance. Aircraft certification regulations (Parts 21–39) are intended to promote the airworthiness of aircraft by requiring that every aircraft, aircraft engine, and propeller is produced in conformance with an approved type design and is in safe operating condition. Aircraft certification regulations also require the development of maintenance requirements and operational limitations.

One of the key goals of certification and continued airworthiness standards is that each safety-critical system have a reliability of at least 0.999999999—"nine 9's"—per flight hour; in other words, the probability that a particular safety-critical system will fail is no more than one in a billion for each flight hour. Regulations seek to achieve this goal through a combination of requirements for design, analysis, test, inspection, maintenance, and operations. As much as possible, regulations do not constrain designers a priori by specifying details such as material properties or the design of individual structures. Instead, designers are given a free hand to incorporate new materials, structural concepts, etc., as long as they accept the responsibility for showing that systems with innovative design features meet the FAA's stringent reliability requirements.

The maintenance rules of Part 43 establish performance standards for individual maintenance workers, and Part 145 establishes quality control system requirements for certificated and rated repair stations that perform maintenance.

Operating rules (Parts 91, 121, 125, and 135) define the responsibilities of owners and operators for ensuring that aircraft are properly operated and maintained in an airworthy condition. Part 119 defines when these operating rules are applicable. Different standards are applied depending on the size and use of aircraft. For example, expected levels of safety for aircraft operated by air carriers is higher than for general aviation aircraft. Neither the "applicability" requirements of Part 23 nor Part 25 addresses the issue of operating rules, which may require that additional equipment be installed or may apply additional airworthiness standards or operating limitations.

Descriptors of aircraft categories used for the purposes of type certification[2] do not directly correspond to similar terms that might be used to define applicable operating rules. The descriptors used in the aircraft certification regulations only define the airworthiness standards applicable to the design of the aircraft, whereas the descriptors used in the operating rules define the specific operating rules that apply. For example, an aircraft type certificated as a "transport category airplane" under Part 25 may or may not be intended for use in air transportation service by an "air carrier" under Part 121. Similarly, an aircraft type certificated as a "commuter category airplane" under Part 23 may or may not be intended for use in "commuter" service under Part 135.

[2]Type certification basis is described below in the section on initial type certification.

Environmental considerations were added to the aircraft certification process by the Noise Control Act of 1972 and by a 1973 amendment to the Clean Air Act. The FAA's Office of Environment and Energy, in close coordination with the Environmental Protection Agency, promulgates the regulations and policies for aircraft noise and engine emissions as required by these statutes. However, findings of compliance are made by AIR as an integral element of the basic aircraft and aircraft engine design approval (type certification) process.

To summarize, the manufacturer is responsible for the original airworthiness of an aircraft, aircraft engine, or propeller; the operator is primarily responsible for maintaining the expected level of airworthiness. The operator retains full responsibility for the airworthiness of its aircraft, even if the operator uses leased aircraft or relies on outside contractors for maintenance. The aircraft certification regulations (Parts 21–39) are developed and administered by AIR, and the operating regulations (Parts 91–135) are developed and administered by the Flight Standards Service. The aircraft maintenance regulations (Parts 43 and 145) are also developed and administered by the Flight Standards Service. However, during the type certification process, the Flight Standards Service works closely with AIR to assess the operational acceptability of the product being certificated and to develop operational and maintenance documentation to facilitate the introduction of the product into service.

Figure 2-1 illustrates how aircraft certification and continued airworthiness regulations are related to the current safety management process.

Supreme Court Opinions Concerning FAA Statutory Responsibilities

In June 1984, the U.S. Supreme Court rendered a landmark opinion concerning the FAA's tort liabilities in the certification of civil aircraft. The court's opinion concerned two cases involving FAA regulatory responsibilities under the Federal Aviation Act of 1958 and determined the extent of the FAA's liability in cases of alleged negligence.[3] The first case concerned the original type certification of a U.S.-manufactured aircraft that was operated by a foreign operator; the second case concerned FAA supplemental type certification of modifications to a small aircraft that had been manufactured in another country. In the second case, the aircraft was used in the United States for air taxi operations

[3]United States v. S.A. Empresa de Vaicao Aerea Rio Grandense (Varig Airlines) et al. (82-1349) and United States v. United Scottish Insurance Co. et al. (82-1350). On writs of certiorari to the U.S. Court of Appeals (9th Circuit 1984). The two cases were combined, and a single opinion was issued.

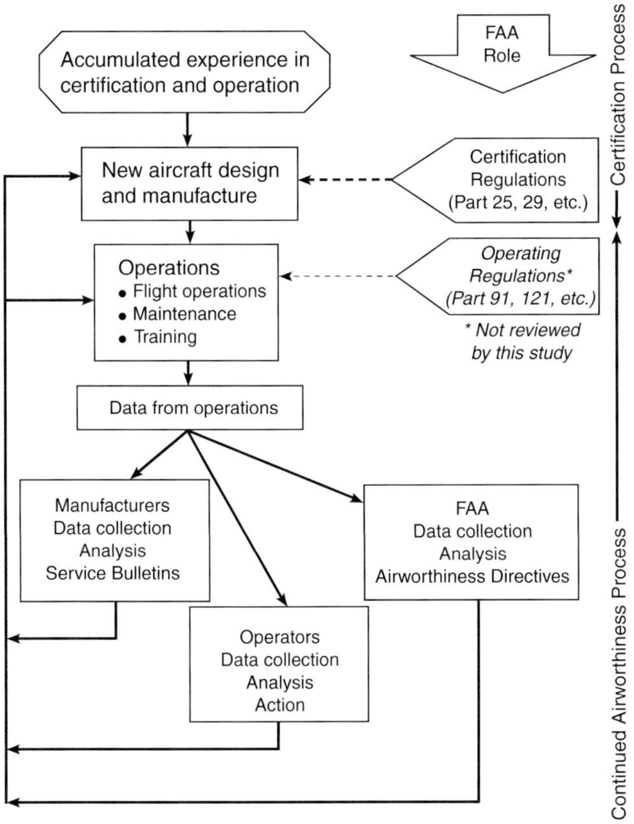

FIGURE 2-1 Current aircraft certification and safety management process. The safety management process recommended by the committee is shown in Figure 4-1.

(i.e., in accordance with Part 135 of the FARs). The Supreme Court determined that liability claims against the FAA for alleged negligence were barred by the "discretionary function exception" of the Federal Tort Claims Act.[4] Important points in the Supreme Court's opinion, which is applicable to all FAA regulatory programs, include the following:

- "The FAA certification process is founded upon a relatively simple notion: the duty to ensure that an aircraft conforms to FAA safety regulations lies with the manufacturer and operator, while the FAA retains the responsibility for policing compliance . . . . This premise finds ample support in the statute and regulations."
- "When an agency determines the extent to which it will supervise the safety procedures of private individuals, it is exercising discretionary regulatory authority of the most basic kind. Decisions as to the manner of enforcing regulations directly affect the feasibility and practicality of the Government's regulatory program; such decisions require the agency to establish priorities for the accomplishment of its policy objectives by balancing the objectives sought to be obtained against such practical considerations as staffing and funding."
- "It follows that the acts of FAA employees in executing the 'spot-check' program in accordance with agency directives are protected by the discretionary function exception as well."
- "The FAA has a statutory duty to *promote* safety in air transportation, not to ensure it." (Emphasis in the original.)

In summary, the court's opinion clarified that the FAA's authority and duty to promote aviation safety through regulation is defined by FARs, whereas the degree to which the FAA exercises oversight authority, especially in areas such as aircraft certification and continued airworthiness, depends to a great extent on the resources allocated by Congress in annual appropriations.

ORGANIZATION OF THE AIRCRAFT CERTIFICATION SERVICE (AIR)

AIR is an organizational unit of the FAA that reports to the associate administrator for regulation and certification, who in turn reports to the FAA administrator. As shown in Figure 2-2, four other units report to the same associate administrator: the Flight Standards Service, the Office of Aviation Medicine, the Office of Rulemaking, and the Office of Accident Investigation.

AIR has a matrix organizational structure with lines of policy direction and guidance that, in some cases, differ from those of the administrative organization. The lines of both policy direction and administrative direction begin with the director of the AIR, who is located at FAA headquarters in Washington, D.C. Technical policy responsibilities within AIR are divided among six "policy centers" (two divisions located at FAA headquarters in Washington, D.C., and four directorates located elsewhere). Each policy center is responsible for regulatory and policy development and for national oversight of its assigned area of technical responsibility. The Aircraft Engineering Division at FAA headquarters is the policy center for general type certification procedures and continued airworthiness procedures, including technical issues common to all aircraft and product types. The Production and Airworthiness Certification Division is the policy center for production and airworthiness certification.

AIR's four directorates are as follows:

- Transport Airplane Directorate (Seattle, Washington)
- Small Airplane Directorate (Kansas City, Missouri)
- Rotorcraft Directorate (Fort Worth, Texas)
- Engine and Propeller Directorate (Burlington, Massachusetts)

Each directorate is also assigned a geographic area that covers about one-fourth of the United States and designated areas overseas. Within its assigned area, each directorate is

[4]Neither case went to trial as a result of the Supreme Court's opinion. Therefore, courts have never decided whether there was negligence on the part of the FAA in these cases.

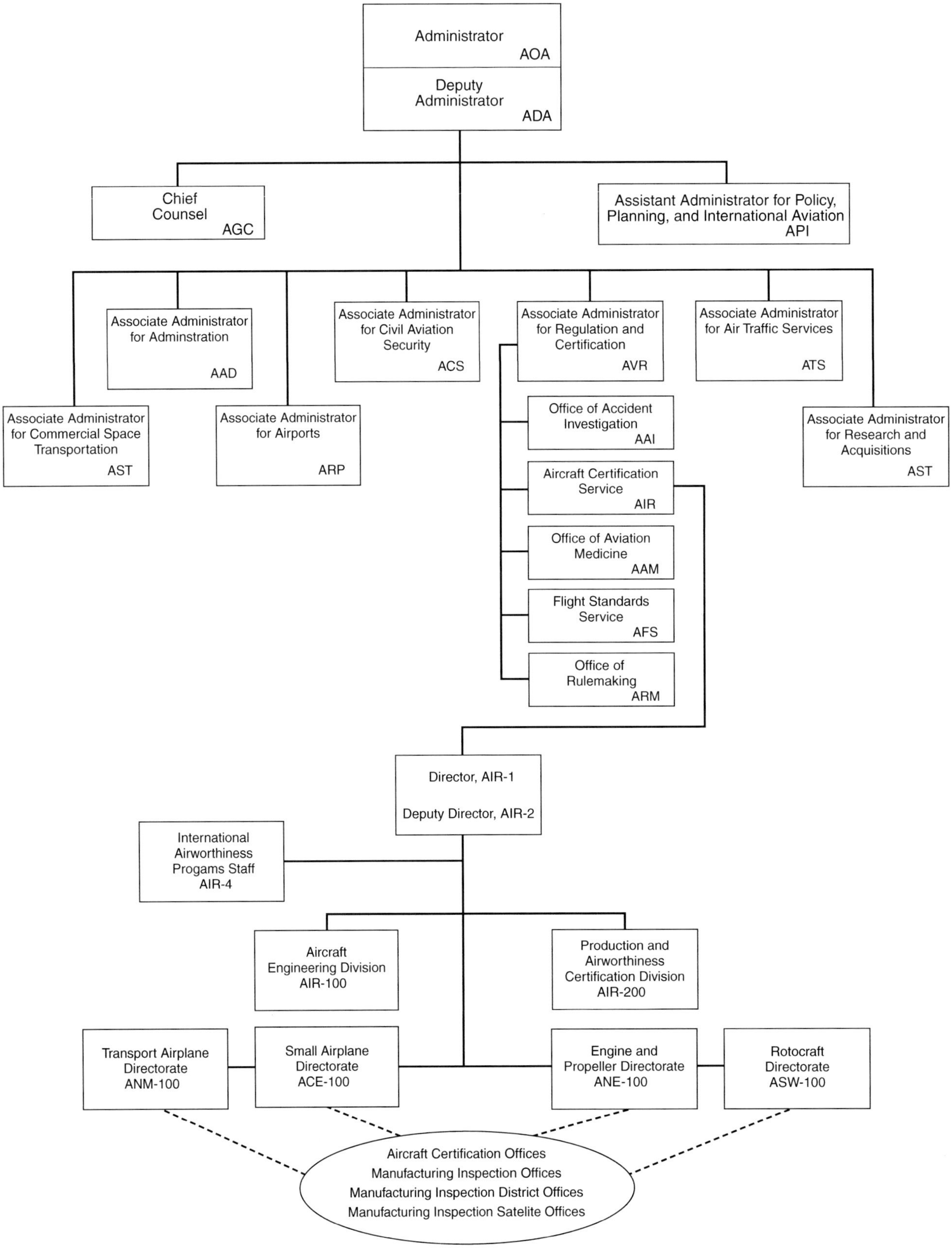

FIGURE 2-2 Partial organizational diagram of the FAA and AIR. Source: FAA.

responsible for all administrative aspects of the aircraft certification and continued airworthiness process. For example, the Transport Airplane Directorate in Seattle is the primary point of contact for all six AIR divisions and directorates for domestic organizations located in 18 western states and for foreign organizations located anywhere in the Asia-Pacific region.

Within each directorate's geographic area of responsibility is an infrastructure of aircraft certification offices (ACOs) and manufacturing inspection district offices (MIDOs), to which the majority of AIR's engineers, flight test pilots, and manufacturing inspectors are assigned. Each ACO and MIDO has a defined geographic area of responsibility within its directorate's geographic region. The ACOs work directly with applicants for design approvals for all types of new aircraft, engines, and propellers; aircraft and system modifications; new materials; and spare parts. Similarly, MIDOs work directly with applicants for and holders of production certificates (manufacturers) for all types of regulated products, primarily with regard to the approval and oversight of production quality control systems. MIDOs are also responsible for issuing airworthiness certificates for individual new aircraft and for approving the airworthiness of engines, propellers, spare parts, etc.

Both the ACOs and the MIDOs have continued operational safety functions that involve reviewing service difficulty reports, participating in accident and incident investigations, developing draft ADs, and enforcing regulations. The ACOs and MIDOs work under the administrative supervision of the directorate with responsibility for their geographic area and receive policy direction and guidance from all other policy centers regarding the product types and technology areas for which each center is responsible. Making this system work effectively requires that all six policy centers work in close cooperation with each other, and AIR has established the Aircraft Certification Management Team for this purpose.

AIR employs a cadre of national resource specialists as technical specialists in key disciplines, such as flight loads and aeroelasticity, nondestructive evaluation, flight management, and human factors. These positions were established to increase AIR's technical expertise in designated areas; national resource specialists work, as needed, on projects throughout AIR.

In accordance with FAR Part 183, AIR also makes extensive use of highly knowledgeable industry personnel designated as "representatives of the administrator." Delegated individuals include designated engineering representatives, designated airworthiness representatives, and designated manufacturing inspection representatives. The FAA relies on these individuals to act on its behalf in reviewing and approving specified actions proposed by their companies. These representatives allow FAA engineers and inspectors to devote more of their attention to more critical areas. Organizational delegations, which are similar to individual delegations, include holders of delegation option authorizations, holders of designated alteration station authorizations, and organizational designated airworthiness representatives.

SPECIFIC REGULATORY ACTIONS

Airworthiness Certification

Airworthiness certificates, which are issued for individual aircraft, are the cornerstone of AIR's overall certification process. Part 91 prohibits the operation of civil aircraft without an airworthiness certificate, in violation of any term of the applicable airworthiness certificate, or in violation of any applicable FARs. The FAA issues both "standard" and "special" airworthiness certificates. Special airworthiness certificates include primary, restricted, limited, provisional, and experimental airworthiness certificates and may be associated with special flight permits that do not meet the international certification standards of the International Civil Aviation Organization (ICAO) in the Convention on International Civil Aviation, to which the United States is a signatory.[5]

For a civil aircraft to receive an airworthiness certificate, the FAA must determine that the aircraft conforms in detail to an FAA-approved type design and is in safe operating condition. Similar requirements exist for engines, propellers, and certain materials, parts, and equipment installed on certificated aircraft. As part of this process, AIR issues several kinds of type certifications and production approvals, which are discussed below.

Type Certification

Type certification includes FAA approvals for new and modified designs of aircraft, aircraft engines, and propellers. Aircraft engines and propellers are issued type certificates separately from the aircraft on which they are installed. One requirement for type certification of an aircraft design is that the designs of the engine and propeller(s), if applicable, have appropriate type certificates. The type certification process also encompasses design approvals for materials, spare parts, and other equipment to be installed on type-certificated aircraft, aircraft engines, and propellers.

Initial Type Certification

When the FAA receives an application for type certification of the design of a new product (aircraft, engine, or propeller), the first step is to establish the type certification basis for the product. The type certification basis is the body of applicable regulations designated in accordance with the procedural regulations of FAR 21.17. Generally speaking, the type certification basis for a product includes the following:

[5]Commonly known as the Chicago Convention of 1944.

- applicable airworthiness standards (e.g., FAR Part 25 for transport category airplanes) in effect on the date of application
- special conditions that have been developed to address novel and unusual design features of the product that are not adequately covered by the basic airworthiness standards
- standards for fuel venting and emissions (Parts 34 and 36)
- exemptions to the above standards that have been issued in the public interest[6]

The type certification basis may be amended during the type certification process for a number of reasons. However, once the type certificate is issued, the type certification basis is fixed and becomes part of the type certificate.

Throughout the production life of a particular product, manufacturers make design changes to improve performance or producibility. Most of these changes are not intended to improve safety. For example, aircraft modifications may be intended to improve dispatch reliability, to reduce maintenance and operating costs, or to improve passenger comfort. As defined by Part 21, these changes are classified as either "minor" or "major" changes.

The FAA evaluates each minor change against the original type certification basis and the body of relevant ADs and, if appropriate, approves it as an amendment to the type design.[7] Approval is required before products built in accordance with the modified design will be issued an airworthiness certificate or approval.

Major changes usually involve the holder of a type certificate who plans to introduce a derivative model of an existing product. The term "derivative" is not defined by regulation but is traditionally used to denote a major change to a previously approved type design. For example, the Douglas DC-9-30 is one of several derivative aircraft designs based on the original DC-9 type design (the DC-9-10).

The type certification basis for a derivative type design is the type certification basis for the original type design, as modified by special conditions issued to address novel and unusual design features in the derivative design that are not adequately addressed by the original type certification basis. Changes to the original type design mandated by ADs must be built into the design of the derivative model, as applicable. Analyses and tests conducted during type certification of the original type design are not repeated if they remain applicable to the derivative design. At the conclusion of the process, the original type certificate is amended to include the derivative model. This process may be repeated to add any number of derivative models to an original type certificate.[8]

Supplemental Type Certification

Manufacturers frequently recommend that owners and operators modify existing products to improve performance, reliability, safety, etc. Typically, these design changes are developed by the manufacturer and are promulgated as service bulletins. Aircraft modification centers and others also design product improvements, such as new navigation and communications installations, upgraded crew and passenger accommodations, modified propulsion systems, and conversions to transform passenger aircraft to all-cargo configurations (complete with large cargo-loading doors and structural reinforcements). In some cases—most commonly with old aircraft—operators must rely on third parties (someone other than the original manufacturer) to design aircraft modifications either because the manufacturer is no longer in business or because the manufacturer is unable to develop the required safety modifications. All of the changes listed above require FAA approval. This approval takes the form of an STC (supplemental type certificate) when someone other than the holder of the type certificate (the original manufacturer) asks the FAA to approve the design for a major change in a product.

As in the case of a derivative aircraft or other product, the type certification basis for an STC is the product's original type certification basis, with new special conditions, as appropriate. To protect the proprietary rights of the type certificate holder, the FAA does not disclose type certification data owned by the type certificate holder without its permission. This permission is often not granted for competitive reasons or because of concerns about product liability. For example, operators may decide to have existing aircraft modified as an alternative to buying new aircraft from the original manufacturer, or the modifications may be intended to compete with similar modifications available from the original manufacturer. If the original manufacturer does not grant access to a product's basic engineering data, an STC applicant may reverse engineer the product design or conduct independent tests to develop the basic product data to proceed with an STC project. For example, if the original strength or functionality of a component to be modified under an STC is established by independent tests and analyses, the STC applicant may then show that the modified component retains at least the same strength or functionality as the original. However, lack of access to the manufacturer's data

[6]Part 11 requires, among other things, that the petitioner for an exemption explain why it would not adversely affect safety or how the petitioner would provide a level of safety equal to the level provided by the rule from which the exemption is sought. Notifications of petitions for exemptions are published in the Federal Register and are subject to public comment before they are granted or denied.

[7]In most cases, manufacturers' designated engineering representatives approve minor changes to type designs.

[8]See Chapter 4 for additional information on the type certification of derivative aircraft, engines, and propellers.

sometimes prevents applicants from formulating an acceptable STC application.[9]

Other approval documents used for type certification include technical standard order authorizations (TSOAs) for the designs of certain materials, parts, and equipment and parts manufacturing approvals (PMAs) for the designs of spare and replacement parts not otherwise approved under a type certificate, STC, or TSOA.

Production Approvals

The FAA issues several types of production approvals: production certificates, production inspection system letters, PMAs, and TSOAs. Production certificates are issued after the FAA examines a manufacturer's production quality control system to verify that the manufacturer has the ability to produce products that conform to the approved type design and are in a condition for safe operation. Production certificates are issued for specific products, and only for products for which a type certificate has already been approved. Approved production inspection system letters, which are similar to production certificates, may be issued to small manufacturers of aircraft, engines, or propellers in lieu of production certificates. The production quality control systems of manufacturers that hold a TSOA or PMA are approved as an integral part of those authorizations. In other words, TSOAs and PMAs are, in effect, combined design and production quality control system approvals.

For all production approvals, the manufacturer's production quality control system is viewed as incorporating the quality control systems of all its suppliers.[10] Thus, each holder of a production approval is responsible for ensuring that its suppliers at all levels perform in accordance with the details of the holder's FAA-approved production quality control system. In the event of a breakdown in a supplier's quality control system, including a breakdown by a subtier supplier, enforcement action by the FAA would be directed at the production approval holder—not the supplier. Thus, although the FAA inspects suppliers, the primary purpose of those inspections is to verify that holders of production approvals are meeting their own oversight responsibilities.

Promoting Continued Airworthiness

AIR monitors the safety performance of aircraft in service, manufacturers' production quality control systems, and representatives of the administrator (designees) to determine if they are maintaining expected levels of safety and to verify they are complying with the terms of certifications or delegations they hold. In effect, the type certification, production approval, and airworthiness certification processes described above never end. If expected levels of safety are not maintained or compliance problems are found, AIR takes corrective action, usually by issuing an AD or taking regulatory enforcement action.

The FAA uses feedback from manufacturers and operators (including pilots and maintenance personnel), as well as its own safety investigations, to determine when corrective action is needed to prevent or correct unsafe conditions. However, the terms "safe" and "unsafe" are hard to quantify. Therefore, the FAA's own measure of public confidence, often manifested by political pressures, can become a factor in the decision making process. For example, ADs have been issued to restore public confidence even before technical investigations have been completed or an agreement has been reached on the cause of an accident or reported safety problem.

AIR's activities to promote the continued airworthiness of the operational fleet of civil aircraft include the following:

- participating in accident and incident investigations
- reviewing and analyzing reports of in-service difficulties that might reveal the existence of unsafe conditions
- reviewing safety recommendations made by the National Transportation Safety Board (NTSB) that relate to aircraft certification
- reinvestigating design approvals for compliance with design certification standards, as necessary
- auditing production quality control systems to verify continued compliance with the terms of production certificates and approvals
- initiating enforcement actions, which may include civil penalties and/or suspending or revoking certificates
- issuing ADs (see below and the section on Risk Management/Action in Chapter 4)

ADs are issued as amendments to FAR 39. Aircraft owners or operators have primary responsibility for determining if their aircraft are in compliance with all applicable ADs. ADs may specify modifications that must be made within a certain time, special inspections that must be conducted, and/or special operating limitations with which the aircraft must comply. Thus, ADs essentially serve as required changes to previously certificated type designs.

Most ADs refer to the detailed instructions in manufacturers' service bulletins or other documents. However, FAA procedural regulations allow the FAA to issue an AD first and then direct the manufacturer of the affected product to submit design changes for FAA approval.[11] After approval, this information is provided to operators. This procedure has been done in a few cases when a manufacturer, for whatever

[9]See Chapter 4 for additional information on STCs.

[10]Some parts suppliers obtain their own production approvals so they can legally market their products to customers other than the prime manufacturers that have designated them as approved suppliers.

[11]Such an AD would, in objective terms, require correction of a stated problem within a specified time.

reason, has been reluctant to develop corrective design changes and make the necessary service information available to operators. In all cases, operators also have the option of proposing corrective action to the FAA.

If a manufacturer is no longer in business or refuses to propose acceptable corrective action, the burden falls on the operators to propose corrective action for FAA approval. Otherwise, affected aircraft cannot be legally operated after the compliance date specified in the AD. Ultimately, if an unsafe condition is determined to exist and there is no known corrective action, the FAA may revoke the airworthiness certificates of the affected aircraft, thus "grounding" them.

In addition to ADs, the FAA has statutory authority to reexamine already certificated aircraft designs and order amendments, suspensions, or revocations of certificates, as necessary to correct unsafe conditions. These reexaminations are especially important for promoting the continued airworthiness of aircraft that are operated much longer than was originally anticipated by their manufacturers. In addition, ADs may be issued or the operating rules (e.g., FAR Parts 91, 121, 125, and 135) may be amended to require operators to install special equipment or to make other modifications to maintain the safety of their aircraft. Examples of such requirements include the requirement that a wide-body transport airplane be able to sustain a large hole in the side of the fuselage; requirements to modify passenger cabins to improve post-crash survivability; and requirements to install collision avoidance systems, ground proximity warning systems, and windshear warning systems.

The FAA has jurisdiction only over operators of U.S.-registered aircraft (i.e., aircraft included in the U.S. Civil Aircraft Registry). However, the airworthiness authorities of most other countries issue either their own orders based on FAA ADs or permanent regulations that require operators in their jurisdiction to comply with FAA ADs as they are issued.

The FAA issues type certificates for many aircraft that are built by foreign manufactures and operated by U.S. operators. ADs issued by the FAA for these aircraft, like ADs issued for aircraft manufactured in the United States, typically refer to the manufacturer's service bulletins for detailed instructions. In most cases, the airworthiness authority in the country of manufacture has already issued the equivalent of an AD under its own regulatory system. However, to require compliance by U.S. operators of those aircraft, the FAA must issue its own AD—in accordance with the U.S. government's normal rulemaking procedures—because the FAA does not have the statutory authority to delegate its rulemaking responsibilities or otherwise subject parties under its regulatory purview to regulations by foreign airworthiness authorities.

In some cases, service experience indicates either to the FAA or to the holder of a type certificate that changes in the type design would improve the safety of the aircraft, even though an unsafe condition does not exist. In those cases, the holder of the type certificate may submit appropriate design changes for FAA approval. Upon approval, the manufacturer makes relevant service information available to operators. In these cases, ADs are not issued, and compliance with the manufacturer's service instructions is at the discretion of the aircraft owner or operator.

Rulemaking Process[12]

The FAA uses a public rulemaking process to comply with its legislative mandate to develop and promulgate minimum safety standards for civil aviation.[13] The FAA also promulgates procedural regulations for administering the application of those standards. AIR develops and issues advisory circulars and internal directives, including handbooks, orders, and notices, to provide guidance on how to implement its policies and procedures.

The rulemaking process is simple in concept. First, the need for a regulatory change is identified. Next, a proposed rule change is developed and published in the form of a notice of proposed rule making (NPRM). Then, public comments in response to the NPRM are evaluated, and a final rule is written, approved, published, and implemented.

The FAA's rulemaking process is structured to comply with applicable procedural requirements, as specified in numerous statues, White House executive orders, Department of Transportation orders and regulations, and the Convention on International Civil Aviation. Whenever the FAA proposes a new regulation, an amendment to existing regulations, or any other rulemaking action, it must cite a relevant statutory authority. If such authority does not exist, the FAA must seek enabling legislation before proceeding. Legislation may also direct the FAA to promulgate specific regulations. For example, the requirement to install collision avoidance systems in most passenger aircraft was the result of legislation that directed the FAA to make appropriate changes to the FARs.

Rulemaking projects may by triggered externally—by public petitions, NTSB recommendations, executive orders, or congressional statutes. Projects may also be triggered internally—by the Office of the Secretary of Transportation, the FAA administrator, the Office of the Chief Counsel, or individual FAA offices, such as AIR. Based on the subject matter of the proposed rulemaking activity, an "office of primary responsibility" is then designated to determine if a project should be initiated and, if so, whether it will be assigned to the FAA's Aviation Rulemaking Advisory Committee (ARAC) or staffed in house by the responsible office.

---

[12]Rulemaking is also discussed in Chapter 6.

[13]All regulations issued by a federal agency, including the FAA, must be promulgated through a public rulemaking process in accordance with U.S. Code Title 5, Chapter 5, commonly called the Administrative Procedure Act. This act is implemented by the FAA in accordance with FAR Part 11, General Rulemaking Procedures.

Recently approved changes to the rulemaking process specify that a Rulemaking Management Council will participate in this decision and other management functions throughout the rulemaking process.

The ARAC and its industry working groups, which operate in accordance with the Federal Advisory Committee Act, facilitate the rulemaking process. Participants in ARAC working groups come from all elements of the commercial aviation industry and the regulatory authorities of many foreign countries. The ARAC provides a vehicle for private individuals and nongovernment organizations to participate officially in the drafting of NPRMs, reviews of public comments, and the drafting of final rules. NPRMs, which are published in the Federal Register, contain the complete draft of a proposed rule and provide interested parties the opportunity to submit data, opinions, or arguments for or against the rule in writing. (See the section on Rulemaking in Chapter 6 for more information on the ARAC.)

Proposed rules must be assessed in terms of their ability to improve safety and their economic impact, including their impact on international trade. A copy of the cost-benefit analysis must be submitted to the Congress before a new rule can take effect. To show that the economic effects are acceptable, the FAA is required by executive order to compare the industry-wide cost of implementing the action with the value (in dollars) of the human lives that the action is expected to save. The Department of Transportation specifies the value (currently, $2.7 million per life) that the FAA must use. The requirement that the FAA provide economic justification for rulemaking actions could prevent the FAA from implementing some NTSB recommendations because the NTSB is not required to screen its recommendations based on economic impact.

The FAA's Office of Rulemaking reports to the associate administrator for regulation and certification. The directors of AIR and other program offices are responsible for the technical substance of proposed regulations; the Office of the Chief Counsel is responsible for the legality and form of proposed regulations; and the Office of Aviation Policies and Plans (which reports to the assistant administrator of policy and international aviation) is responsible for verifying compliance with executive orders that require cost-benefit or regulatory analyses.

Rulemaking procedures, as authorized by legislative mandate and implemented by FAR Part 11, generally take a long time to complete. The FAA often provides 120 days or more for public comments in response to an NPRM. In addition, final rules must be published in the Federal Register at least 30 days prior to their effective date, and many final rules specify compliance dates beyond the effective date of the rule. However, the FAA may issue a regulation without an NPRM or a 30-day period for public notice if it determines that these would be impractical, unnecessary, or contrary to the public interest—and explains why in the final rule (for example, in the case of an emergency involving safety or security). The preamble of each final rule summarizes all public comments received and explains how and why the FAA reacted to those comments.

The FAA also uses the Airworthiness Concern Coordination Process (better known as the lead airline concept) to obtain operational insight regarding potential airworthiness problems and proposed solutions. This process comes into play when the FAA needs to interact with the Air Transport Association (an association of major U.S. airlines) in developing ADs or other actions relevant to a specific aircraft model. Under this process, which seems to be working well, the Air Transport Association has designated a specific airline for each aircraft model to take the lead in working with the FAA.

AIR works with the JAA (the Joint Aviation Authorities, an organization comprised of European regulatory authorities) and individual regulatory authorities of other foreign countries to promote international harmonization of aircraft certification regulations, policies, and practices. In fact, many rulemaking projects are triggered by efforts to harmonize specific regulations between the United States and Europe.

The authority to issue NPRMs related to aircraft certification has been delegated to the AIR director. With some exceptions, final rules are issued by the FAA administrator. Exceptions include the approval of ADs and special exemptions from the airworthiness standards that would normally apply to the type certification of particular products. The authority to issue final rules in these areas has been delegated to the managers of AIR's four directorates in their areas of responsibility.

Final rules that may have a significant economic impact or may generate significant public (or political) interest must obtain concurrence from the Office of the Secretary of Transportation and the Office of Management and Budget (OMB) before they are issued. In some cases, concurrence is required even before an NPRM can be issued.

Resources and Level of Activity

AIR's budget for fiscal year 1998 was $101 million. At the end of fiscal year 1997, AIR had 1,010 full-time permanent employees, including 678 nonsupervisory technical personnel (aviation safety engineers, aviation safety inspectors, and flight test pilots), 94 managers, and 238 support personnel. This staff was distributed over 34 geographical locations. Activities completed by AIR during fiscal year 1997 included the following:

- 25 type certificates approving the designs of new aircraft, aircraft engines, and propellers
- 110 amendments to existing type certificates approving the designs of new derivative models of aircraft, aircraft engines, and propellers
- 1,011 new STCs approving designs for modifications to products approved by preexisting type certificates

- 530 amendments to existing STCs
- 5,715 lesser design approvals (such as TSOAs and PMAs)
- 2,473 production certificates and other production quality control system approvals
- 1,840 original airworthiness certificates issued to newly manufactured aircraft
- 341 ADs to correct safety problems in existing aircraft
- 790 appointments of new designees to act as "representatives of the administrator" for design approvals and airworthiness certifications
- final action in response to 31 petitions for exemption
- final action in response to 14 petitions for rulemaking

The FAA as a whole published 19 NPRMs and issued 29 final rules.

RELATED ACTIVITIES

Unapproved Parts

Unapproved parts are parts that have not been approved under applicable FARs or have lost their approval status because of a change in condition caused by damage, wear, etc. Unapproved parts are created primarily in the following ways:

- Parts are manufactured by individuals who do not seek FAA approval because they are unaware of relevant requirements.
- Parts from overruns produced by suppliers to prime manufacturers find their way into the marketplace without undergoing the required quality control inspections to determine their conformity or condition.
- Surplus military parts are sold as scrap and then introduced into the marketplace for civil aircraft parts with false documents misrepresenting them as serviceable parts.
- Counterfeit parts are produced by unknown sources and falsely marked, packaged, or otherwise misrepresented as parts manufactured by an approved manufacturer.
- Parts that are approved by foreign governments for installation on aircraft on their registry find their way into the U.S. marketplace without FAA approval for installation on U.S.-registered aircraft.
- Previously approved used parts are reintroduced into the marketplace with bogus, falsified, or altered maintenance records even though the parts have outlived their safe life, have been inappropriately repaired, and/or have been reconditioned to look serviceable even though they are not. This group may include parts taken from aircraft that were involved in accidents.

Thus, an unapproved part may be functionally identical in all respects to an approved part but lack the proper documentation or authorization. Although using this kind of unapproved part does not present a safety problem, per se, it is often difficult to distinguish an undocumented but fully functional part from a close copy that appears to be fully functional but, in fact, is not. Thus, the practice of using unapproved parts degrades safety.

Because many approved parts cost much more than similar unapproved parts, there is a strong financial incentive for dishonest vendors or repair facilities to market or use unapproved parts knowingly. For example, in one well publicized case, an unapproved bearing spacer was detected by a repair facility conducting maintenance on the engines of a large transport airplane. The unapproved spacers, which were commercially available for other applications, could be purchased for about $40, whereas the genuine, approved part cost $500. The unapproved part would disintegrate after 600 hours in flight, resulting in complete engine failure, whereas the approved part had a 20,000 hour replacement life. The FAA issued an AD to inspect engines with the same part, and more than 30 unapproved parts were found on in-service aircraft.

There are documented cases that unapproved parts have caused fatal accidents. One involved a transport category aircraft airplane and resulted in 55 fatalities. That particular accident occurred in Denmark, but the aircraft involved was manufactured in the United States. One of the cause factors for the accident was the use of unapproved and substandard parts, some of which were reportedly supplied by a company located in the United States. In another case, the failure of a helicopter tail rotor in New Zealand caused two fatalities. This accident was also caused by an unapproved part supplied by a U.S. company.

FAA action to police this problem is complicated by the large number of parts suppliers, which are not required to be registered with or approved by the FAA. It is the responsibility of the individuals and companies that perform maintenance to ensure that they use only approved parts, but this can be a difficult task if a parts supplier knowingly provides unapproved parts. To address the risk posed by unapproved parts, the FAA has established an office dedicated to this problem. This office has established procedures applicable to AIR personnel, the Flight Standards Service, and other FAA offices involved in the unapproved parts problem. The FAA has also recognized self-governing approval processes established by two groups, the Society of Automotive Engineers and the Airplane Suppliers Association, to facilitate "accreditation" of qualified parts suppliers that have lacked formal approval.

The committee believes the FAA should continue its current efforts to address the unapproved parts problem, including industry education, investigation, and appropriate regulatory action.

Aging Aircraft

The aging aircraft initiatives carried out by the FAA, overseas authorities, airlines, and manufacturers have been very successful in addressing the structural problems of aging aircraft, which were highlighted by the 1988 accident of an Aloha Airlines Boeing 737. Since then, all major jet transport airplanes have been subject to extensive, mandatory structural inspections and modifications. Aging aircraft programs, however, have not yet fully addressed issues associated with widespread fatigue damage criteria and repairs.

The FAA's aging aircraft program evaluates aircraft structures in accordance with damage tolerance principles. These evaluations, however, do not consider the effects of structural repairs of the airworthiness integrity of the aircraft. The committee believes existing repairs and repair procedures should be reassessed using damage tolerance concepts. The FAA's ARAC has recommended a special FAR that would require operators to review repairs on airplane pressure cabins. In addition, major manufacturers have prepared repair assessment guidelines. The program is under way on a voluntary basis, but regulatory action, which may take several more years to complete, is required to mandate compliance by all operators. The committee believes the FAA should expedite action in this area.

In early 1997, the White House Commission on Aviation Safety and Security recommended expanding the FAA's aging aircraft program to include nonstructural systems. However, there are comprehensive maintenance and overhaul procedures for nonstructural systems, and it is not clear if an aging problem does in fact exist in areas other than the structure. The committee believes it would be worthwhile for the FAA to initiate a cooperative study on aging of nonstructural systems to evaluate the need for expanding aging aircraft initiatives in this area.

REFERENCES

FAA (Federal Aviation Administration). 1998. Aircraft Certification Mission Statement. Federal Aviation Administration. Online. Aircraft Certification Service (AIR) Headquarters Office Home Page. Available: http://www.faa.gov/avr/air/hq/mission.htm. February 10, 1998.

Komons, N.A. 1978. Bonfires to Beacons: Federal Civil Aviation Policy under the Air Commerce Act, 1926–1938. Washington, D.C.: Federal Aviation Administration.

White House Commission on Aviation Safety and Security. Vice President Al Gore, Chairman. February 12, 1997. Final Report to President Clinton. Washington, D.C.: White House Commission on Aviation Safety and Security.

3

Causes of Incidents and Accidents

INTRODUCTION

Aviation safety experts have realized for some time that aircraft incidents and accidents almost always result from a series of events, each of which is associated with one or more cause factors. Thus, the cause of an accident or incident has many aspects. Some internationally accepted definitions in the context of the investigation of an aircraft accident or incident are listed below (ICAO, 1994):

- **Causes** are actions, omissions, events, conditions, or a combination thereof, that lead to an accident or incident.
- **Accidents** are occurrences associated with the operation of aircraft, from the time any person boards an aircraft with the intention of flight until the time all persons have disembarked, that results in one or more of the following:
  - —A person is fatally or seriously injured.
  - —The aircraft sustains damage or structural failure that adversely affects the structural strength, performance, or flight characteristics of the aircraft and would normally require major repair or replacement of the affected component.
  - —The aircraft is missing or completely inaccessible.
- **Incidents** are occurrences, other than accidents, associated with the operation of aircraft that affect or could affect the safety of operation.

The definition of cause given above takes into account the many events involved in an accident or incident. These events can be viewed as links in a chain. Investigations of some hull loss accidents in the United States have revealed as many as 20 links in the chain; the average is just under 4 links.[1] For example, after an exhaustive technical and legal investigation into one controlled flight into terrain (CFIT) accident, an official commission concluded that at least 10 essential cause factors were involved.[2] If any one of these 10 cause factors had not been present, or if some of the factors had occurred in a different order, the accident would not have happened. The most effective accident prevention strategy must take into account all the links in the chain of events that lead to incidents and accidents.

Subdividing an incident or accident into a chain of events reveals important information. If one more element is added to the chain in an incident, for example, the consequences of the incident might be much more serious, even resulting in an accident. Conversely, removing one link in the accident chain could substantially mitigate the consequences or, possibly, prevent all adverse consequences. In other words, from a safety management viewpoint the only meaningful difference between many incidents and accidents is the consequences.

For example, an aircraft may experience several abnormalities involving equipment malfunction, unexpected adverse weather conditions, and loss of situational awareness by the flight crew. As a result, the aircraft may take longer than expected to slow down after landing. If the aircraft happens to be landing at an airport with runways of the minimum required length with water hazards at the end, there could be a catastrophe. The resulting investigation might lead to a comprehensive review of procedures and systems related to approach and landing. If the same sequence of events happened at an airport with runways of the minimum required length but with a *grassy field* at the end, the aircraft might run off the end of the runway and experience minor damage and no crew or passenger injuries. In that case, there

[1]Accident data in this chapter are primarily related to hull loss accidents. For the purposes of this study, data from all accidents and from fatal accidents are not significantly different from data from hull loss accidents in terms of causes and trends in the accident rate.

[2]A CFIT accident occurs when a mechanically sound aircraft collides with the ground, typically because the flight crew loses situational awareness and does not understand the flight path of the aircraft relative to the ground. "Loss of control" accidents include collisions caused by engine failure, icing, stalls, or other circumstances that interfere with the ability of the flight crew to direct the motion of the aircraft.

would be a review of the incident within the aviation community but little public notice. *But if the same sequence of events* happened at an airport with runways that were significantly longer than required for that aircraft to land, the aircraft might still stop well short of the end of the runway. In that case, depending upon the attentiveness of air traffic controllers and the inclination of the flight crew, a report might not be filed, and there might be no examination of how to prevent a similar series of events from happening in the future. In that case, the opportunity to take proactive safety measures before an accident happens would be lost. The challenge of aircraft safety management is identifying and focusing attention on truly hazardous conditions *before* a potential accident becomes a reality. In the example described above, routine use of flight recorders or quick access recorders (QARs) to monitor stopping distance would provide operators with an independent means of detecting potentially hazardous abnormalities.

In 1996, accidents involving jet transport airplanes occurred in the United States at the rate of about 1.46 accidents per million flights (or one accident for every 685,000 flights). Except for a few, well publicized tragedies, most passengers involved in commercial aircraft accidents are not killed or injured. However, as shown in Figure 3-1, the rate for hull loss accidents involving large jet transports has improved only slightly in the last 20 years. The same is true for fatal accidents.

Accident rates can be computed in terms of accidents per passenger-trip, accidents per passenger-mile, or accidents per passenger-flight hour. Depending upon the method chosen, the accident rates for Part 135 operators, who commonly use turboprop aircraft, are about three to eight times higher than the accident rates for Part 121 operators, who operate most of the large jet transports (FAA, 1994). Rates for Part 135 operators are declining, but slowly. Regulatory changes to establish more uniform safety standards for Part 121 and 135 operators are intended to address the disparity in accident rates.

Analyses of the chains of events in accidents are generally useful just for preventing similar accidents. Because there are so few accidents in the United States relative to the number of flights, focusing safety programs on accidents alone addresses only a small fraction of potential accidents and is reactive rather than proactive. A proactive approach that could eliminate risks before they cause accidents requires an effective means of tracking the chains of events in both incidents and accidents. Preventive action (not just remedial action) could then be taken—based on how often individual links in the chain recur and their potential for contributing to future incidents and accidents. Every abnormal event in the incident or accident chain could be examined to identify the cause factors that explain why it happened and to describe the underlying problems and deficiencies that should be corrected.

One approach for visualizing an incident or accident is as a chain of events that must occur in a certain sequence. Another is shown in Figure 3-2, which shows a system of disks spinning at random. Each disk contains a hole that must line up precisely with the holes in the other disks before a beam of light can pass through the entire system. The probability

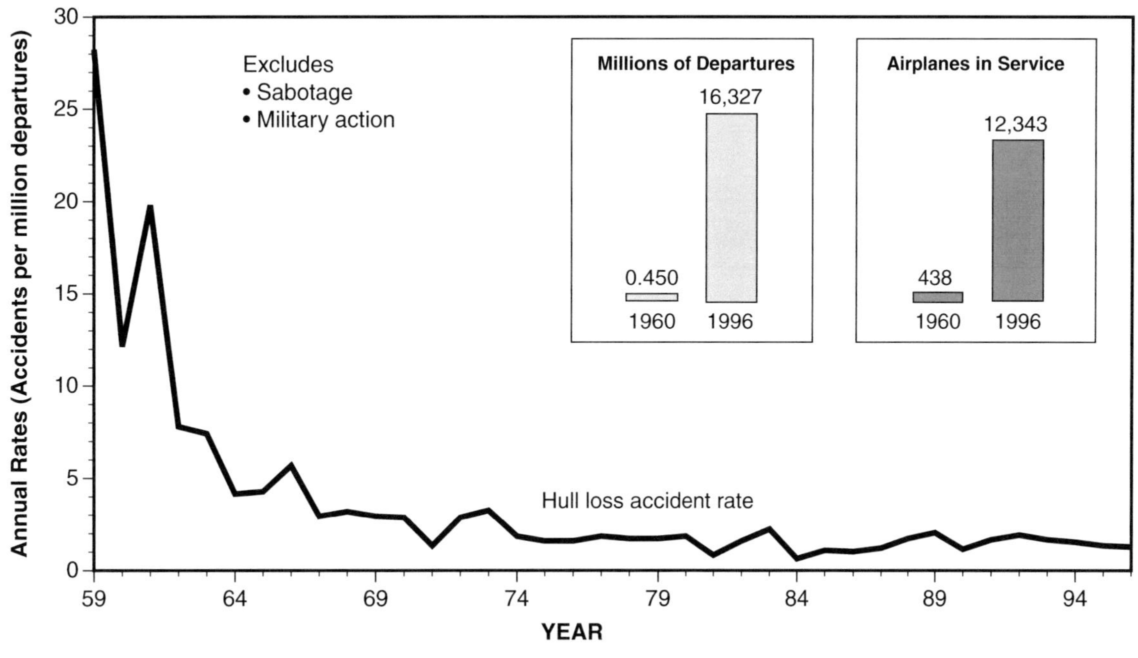

FIGURE 3-1 Worldwide hull loss accident rates, 1959 through 1996. The data depicted in all Chapter 3 figures cover commercial jet aircraft heavier than 60,000 pounds. Available data on aircraft manufactured by the states of the former Soviet Union are incomplete and are, therefore, excluded. Data on accidents caused by sabotage, hijacking, military action, experimental test flying, or suicide are also excluded.

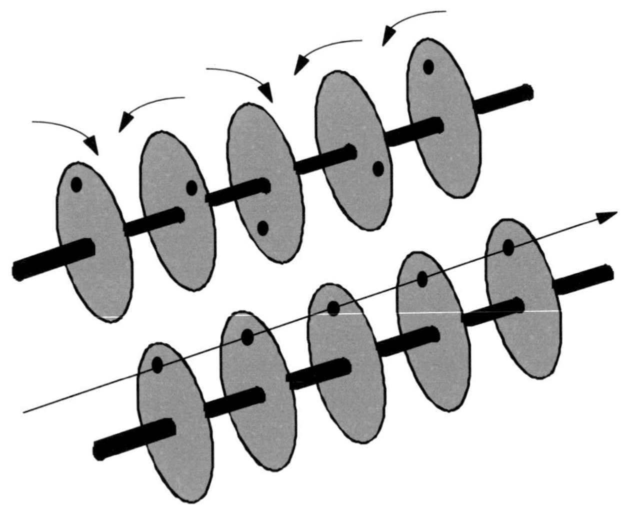

FIGURE 3-2 "Spinning disk" view of accident and incident events. Source: Reason Model (Reason, 1990).

that the disks will all line up after being spun "independently" is less than if the positions of two or more disks are linked. This is why primary and backup hydraulic systems are physically separated as much as possible—so that a single damaging event will not disable all hydraulic systems. Assuming that the disks represent events in a chain leading to an accident, corrective action to prevent one or more of the events (i.e., filling in the holes) is one approach to preventing other accidents that might involve the same events.

When an official investigator reports the "probable causes" of an accident or incident, consideration should be given to all of the events and cause factors. Cause factors can be grouped into the following categories:

- human factors/personnel error
- malfunction or failure of aircraft structures, engines, or other systems
- deficient maintenance
- hazardous environment involving weather, volcanic ash, birds, etc.
- air traffic management errors
- any combination of the above

Identifying the precise cause factors for each event can be complicated, requiring good judgment and accurate interpretation of the facts. There could be more than one cause factor for each event, and some cause factors naturally overlap.

Human factors include mistakes caused by voluntary acts, failure to act, and other factors associated with actions or inaction.[3]

Cause factors associated with aircraft, engines, and systems include deficiencies in the design, manufacture, maintenance, or operation of the aircraft or its systems.

[3]An expanded list of cause factors appears in Appendix D.

Maintenance-related cause factors include improperly performed maintenance and inadequate maintenance procedures and plans.

Environmental cause factors include hazardous weather, volcanic ash, sand, dust, and birds.

Cause factors associated with air traffic management include deficiencies in weather reporting, regulations, and the air traffic control system (navigational aids; air traffic control directives; and airport facilities, runways, and taxiways).

Combinations of factors and cascading cause-and-effect sequences must be carefully examined to understand all of the cause factors. For example, to prevent accidents caused by system failure, the system that failed could be modified to prevent similar failures in the future. In addition, understanding if the failure was triggered by the failure of some other system, improper maintenance, abnormal operating environment, etc., may suggest additional corrective action.

PRIMARY CAUSES

The term "primary cause," defined as the most critical cause factor associated with a particular incident or accident, can be deceiving and is often subject to interpretation. One cause factor may contribute more to the consequences of an accident or incident than the others, but making this determination may also depend on one's point of view. Take the case of an accident involving an uncontained engine failure that severed all of the aircraft's hydraulic lines.[4] The uncontained engine failure disabled the hydraulic systems needed for conventional flight control; the aircraft was controlled by varying the thrust asymmetrically on the remaining engines until the plane was just short of a runway on which the aircraft was attempting to make an emergency landing. The aircraft crash landed, broke apart, and caught fire. The flight crew and some of the passengers survived the accident.

The official investigation found that the original material from which a large rotating part of the engine was fabricated contained a defect that ultimately resulted in a crack. The crack grew over the life of the part and finally fractured, resulting in shrapnel damage to the aircraft and its hydraulic systems. The investigation further disclosed that the part had undergone numerous inspections designed to locate defects like the one that ultimately resulted in the part failure. Inspections were made at the part's material manufacturer, the forging manufacturer, the engine manufacturer, and during routine maintenance of the engine by the operator. The official report on this accident determined that the probable cause was inadequate consideration given to human factors limitations in the inspection and quality control procedures

[4]In an uncontained engine failure, a piece of the engine, such as a rotor disk, is ejected from the engine. Commercial jet engines are designed to contain blade failures but not disk failures. The danger of disk failures is, therefore, addressed through stringent manufacturing and inspection procedures.

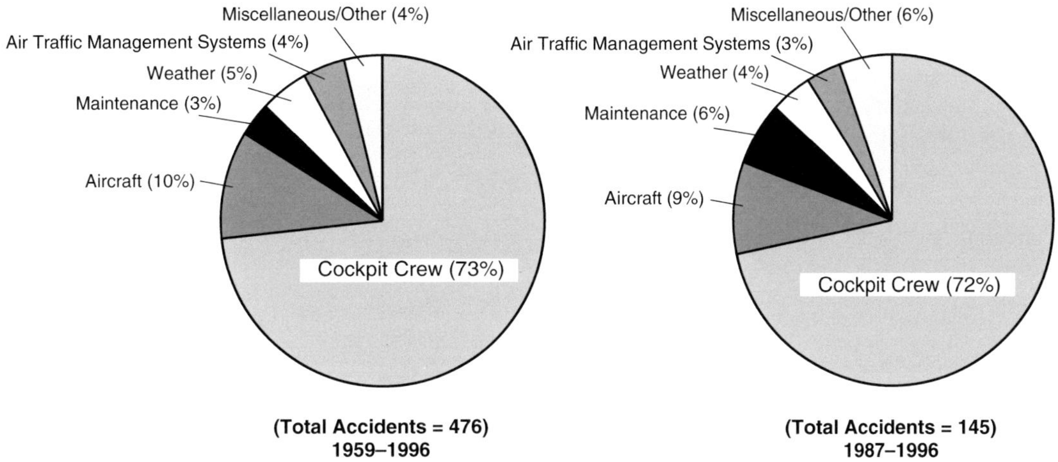

FIGURE 3-3 Primary cause factors for hull loss accidents involving large commercial jet airplanes worldwide. Source: Boeing, 1997.

used by the airline's engine overhaul facility. As a result, a fatigue crack originating from a previously undetected metallurgical defect was not detected. The subsequent catastrophic disintegration of the part produced debris with a pattern of distribution and with energy levels that exceeded the level of protection provided by design features of the hydraulic systems that operated the aircraft's flight controls.

In this example, one link in the chain was singled out as being more significant than the others, which included the processes used to produce the basic material for the rotating part, numerous inspections designed to detect the defect before it became a crack, and the susceptibility of the aircraft design to damage by the distribution of debris in this particular failure. The committee believes that a safety management program should have an inclusive view of what constitutes a significant cause to ensure that corrective action addresses multiple cause factors and provides multiple assurances that a similar accident or incident will not occur in the future.

CAUSES OF JET TRANSPORT ACCIDENTS

For the purposes of analyzing the most common causes of accidents, the committee reviewed official accident reports and various summaries of those reports. Figures 3-3 and 3-4 show the primary cause factors cited in official reports of accidents resulting in the loss of aircraft. Data from both the U.S. and worldwide fleets of commercial jet airplanes are shown for two periods: since the beginning of the "jet age" (1959 to 1996) and during the last 10 years (1987 to 1996) (Boeing, 1997).

Figure 3-3 shows that there were 50 hull loss accidents worldwide between 1959 and 1996 in which the airplane was a primary cause factor. Figure 3-4 shows that 15 of these accidents took place in the United States. For the 50 accidents worldwide, Figure 3-5 shows the breakdown by aircraft system.

Although data on primary cause factors are readily

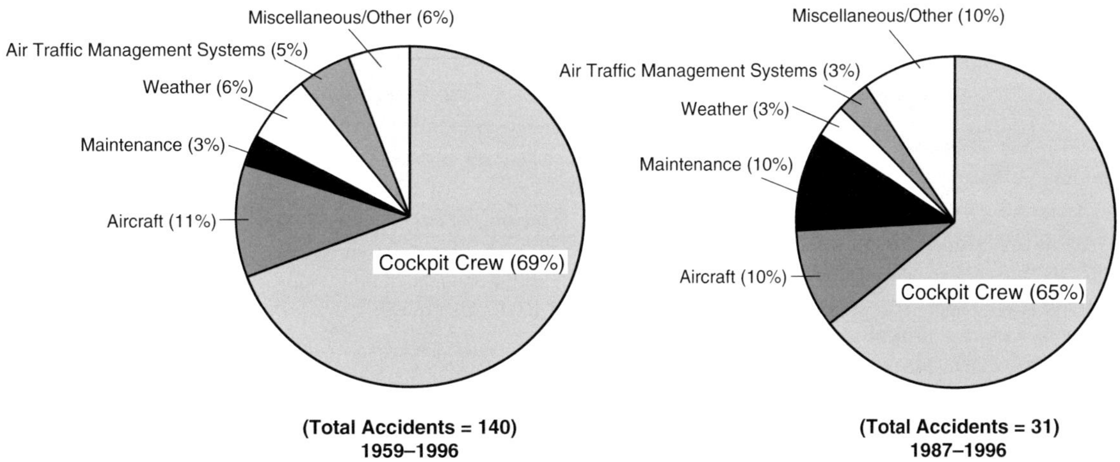

FIGURE 3-4 Primary cause factors for U.S. hull loss accidents involving large U.S.-registered commercial jet airplanes. Source: Boeing 1997.

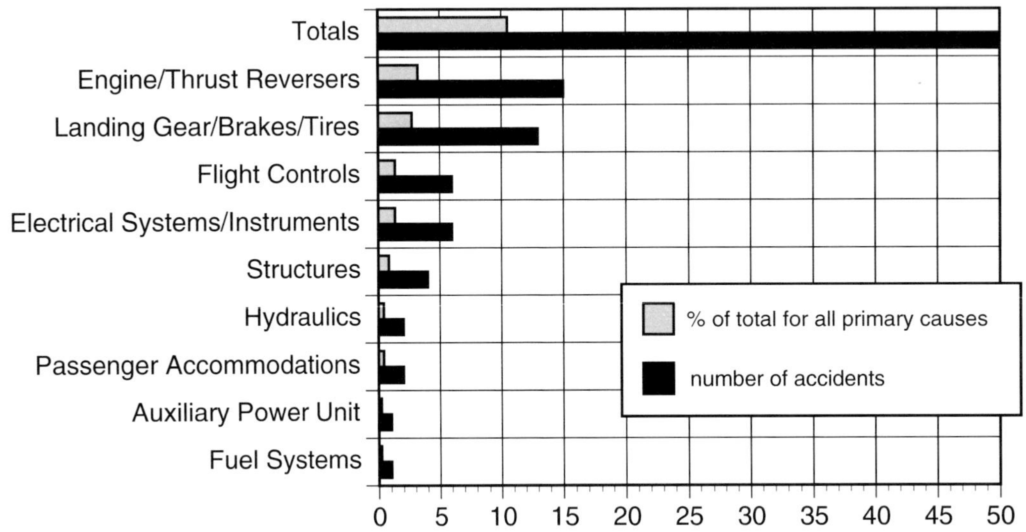

FIGURE 3-5 Airplane system cause factors for hull loss accidents involving large commercial jet airplanes worldwide, 1959 through 1996. Source: Boeing, 1997.

available, these data are often misleading when used to identify detailed trends in accident causes. For example, Figure 3-2 might lead one to believe that maintenance is a growing problem. The percentage of maintenance-related accidents during the 10-year period ending with 1996 is more than three times higher than for the entire 38-year period (1959 to 1996). However, when dealing with small numbers, small changes can produce large changes in percentages. In fact, only three accidents during the 10-year period were attributed primarily to maintenance. As already discussed, the attribution of primary causes is sometimes problematic. If the primary cause of two of those accidents had been attributed to one of the other cause factors associated with those accidents, there would have been no percentage increase in maintenance-related accidents.

One manufacturer examined a large number of past accidents and identified all actions that could have broken the chain of events leading up to the accidents. This examination indicated that operators could significantly reduce accident risks by taking the following measures:

- implementing a comprehensive flight operations quality assurance (FOQA) program such as the British Airways Safety Information System (BASIS) to monitor adherence to standard operating procedures and identify operational irregularities that could foreshadow accidents and incidents[5]
- establishing training programs that emphasize basic piloting skills, upset recovery techniques, cockpit discipline, the use of standard operating procedures, and crew coordination and crew resource management

[5]This report uses FOQA as a generic term that is not limited to the specific program of the same name that the FAA sponsors. Chapter 4 discusses FOQA and BASIS in more detail.

An accident prevention strategy that considers all cause factors involved in incidents and accidents—not just primary cause factors—has a greater potential to prevent accidents by eliminating factors that are common in many incidents and accidents. These common factors serve as "traps" that may be easier to identify and eliminate than a unique, extremely rare factor that may be labeled the "primary cause" in a given accident. For example, if a series of accidents appears to be unrelated, corrective action might focus on the specific circumstances of each accident. A comprehensive review, however, might reveal a fundamental deficiency, such as poor pilot training, safety management, or aircraft maintenance, that is common to the entire series of accidents. Identifying and correcting these fundamental deficiencies is important because they can lead to many types of incidents and accidents.

Trend analysis based on reliability, or mean time between failure, could add another dimension to the safety management process. One could theoretically do trend analyses of aircraft components, structures, etc., to keep faulty parts from becoming causal in the chain of events leading up to an incident or accident, and this would enhance safety. However, because of the redundancies built into the design of aircraft structures and systems, the failure of any single component does not pose a threat to continued safe operation. In fact, FAA-approved minimum equipment lists allow aircraft operation with some equipment out of commission. Also, for economic reasons, airlines and manufacturers already use component reliability analyses to keep their aircraft in the air. For at least the timeframe of this study—the next 10 years—the committee believes that a focused effort to determine mean times between failure, which would require collecting and analyzing vast amounts of data, might not identify specific safety trends and would bog down the safety management process, making it more difficult to identify specific, cost-effective preventative action.

CAUSES OF JET TRANSPORT INCIDENTS

Data on incidents involving jet transport airplanes provide a slightly different picture. To begin with, many organizations do not have adequate incident reporting systems, and it is very difficult to obtain complete and consistent records of incidents. Whereas accidents tend to be highly visible, are consistently reported, and are carefully investigated, incidents include a broader range of situations and cause factors, are so numerous that available resources in industry and government are insufficient to conduct thorough investigations of most reported incidents, and reporting them often depends on the initiative of the personnel involved (who may have a conflict of interest if the report is likely to have negative consequences for them). In addition, broadly accepted definitions of what constitutes an incident are imprecise and, in practical settings, they are interpreted differently by different organizations and individuals.

Table 3-1 shows data resulting from an examination of 2,032 incidents worldwide that were reported over a 10-year period for aircraft built by a particular manufacturer. The aircraft included in this examination accounted for about one-fourth of the world's large transport airplanes. The reader should keep in mind that manufacturers have a special interest in preventing incidents and accidents associated with system malfunction. Therefore, a jet transport manufacturer's database may be biased toward incidents in which aircraft system performance is involved. Wherever possible, each incident was broken down into a sequence of events. A total of 1,618 events were identified and categorized by the links in the chain of events and their cause factors. Table 3-1 shows the number and percentage of the cause factors associated with each event. Figure 3-6 shows a breakdown of all cause factors for all events by aircraft system. This analysis gave equal emphasis to all factors in the chains of events.

TABLE 3-1 Causes of Aircraft Incidents

Cause Factor	Number of Events	Percentage
Personnel (human factors)	800	49.44
Aircraft	547	33.81
Maintenance	214	13.23
Environment	33	2.04
Air traffic management	24	1.48
Totals	1,618	100.00

Source: Boeing.

SUMMARY

Because accidents are rare, analyses of accident records can provide guidance on broad areas of concern but are inherently incapable of preventing other types of accidents. Incidents are more frequent and are a rich source of safety data, but the quantity of the data is so large that it is difficult to identify meaningful risks and avoid unfruitful diversions. The process is complicated because some accidents are truly unique and may not be indicative of future hazards, whereas some seemingly inconsequential incidents are disasters waiting to happen.

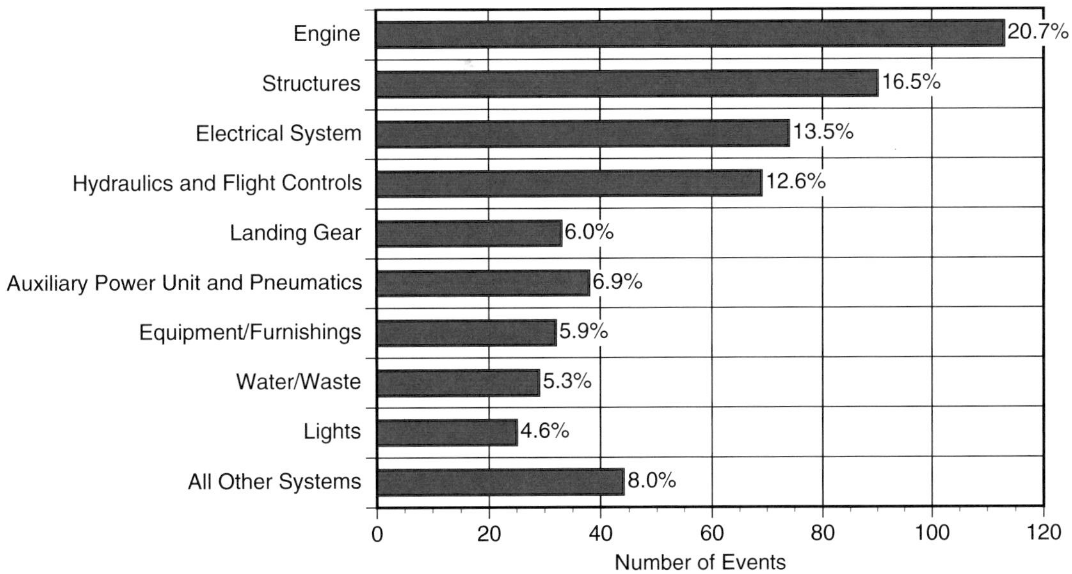

FIGURE 3-6 Airplane-related cause factors in worldwide incidents involving large commercial jet aircraft produced by a particular manufacturer (about 25 percent of the worldwide commercial jet fleet), 1987 through 1996. Source: Boeing.

Accidents and serious incidents almost always have multiple causes, although many analyses and safety records focus on "primary" causes. This narrow focus diverts attention from other cause factors that were essential links in the chain of events and that should also stimulate corrective action to prevent future accidents. With careful analysis, however, a safety management process can identify accident prevention strategies that eliminate factors ("traps") that recur in many different accidents. Such a process could effectively reduce many different types of accidents by eliminating the co-factors necessary for their occurrence.

Personnel error (human factors) is the most common cause of both incidents and accidents. CFIT and loss-of-control accidents, which almost by definition involve human factors, account for more than half of all fatal accidents. Similarly, inappropriate crew response and fuel exhaustion, which are also essentially human factors problems, are the major contributors to propulsion-related fatal accidents. Although aircraft system malfunctions are involved in a relatively small fraction of aircraft incidents and accidents, improvements in aircraft systems often improve safety by making aircraft more robust—providing flight crews with more accurate information to improve their situational awareness and reducing the likelihood that a human error will result in an incident or accident.

Finding 3-1. Safety management processes that focus on the primary causes of accidents are reactive and are unlikely to address some important cause factors adequately. Data from investigations of accidents *and incidents* are essential for planning proactive corrective action, which should address all important cause factors.

REFERENCES

Boeing. 1997. Statistical Summary of Commercial Jet Airplane Accidents, Worldwide Operations 1959–1996: Airplane Safety Engineering. Seattle, Wash.: Boeing Commercial Airplane Group.

FAA (Federal Aviation Administration). 1994. 1994 Statistical Handbook of Aviation. Washington, D.C.: FAA.

ICAO (International Civil Aviation Organization). 1994. International Standards and Recommended Practices: Aircraft Accident and Incident Investigation. Annex 13 to the Convention on International Civil Aviation, 8th ed. Montreal, Canada: ICAO.

Reason, J. 1990. Human Error. Cambridge, U.K.: Cambridge University Press.

4

Recommended Safety Management Process

OVERVIEW

The current safety management process, which has evolved over the history of the aviation industry (see Figure 2-1), involves operators, manufacturers, and the FAA. Despite an outstanding safety record for the commercial aviation industry, the accident rate, as a function of departures, has improved only slightly in the last 20 years. At the same time, aviation travel (revenue passenger miles) has been increasing steadily (around 5 percent per year) thus increasing risk exposure. Public expectations regarding safety have also been increasing. It has become apparent that a significant, continuing reduction in the already low accident rate will require improvements in all aspects of the aviation industry, including the aircraft certification safety management process.

The Major Finding. The recommended safety management process should improve the ability of the FAA/AIR, manufacturers, and operators to take corrective action based on incident data—before an accident takes place—and to set priorities based on assessments of current and future risk. However, the current process is already highly effective—as indicated by the small contribution of aircraft system malfunctions to the overall accident rate—and changes to the current system must be carefully structured to avoid unintended consequences that might reduce safety in some situations.

Priorities

The vast majority of aircraft that will operate during the next 10 years either have already been manufactured or will be manufactured to design specifications that have already been certificated. Therefore, monitoring the safety of operating aircraft is essential to obtaining an accurate understanding of safety levels, to detecting and resolving problems as soon as possible, and to validating airworthiness standards. Historically, improvements in standards for initial type certification have frequently been based on lessons learned from the continued airworthiness process. Consequently, making the continued airworthiness process more effective is essential to improving safety in the near term and to providing a foundation for improvements in the long term. Even though the primacy of continued airworthiness is reflected in AIR's mission priorities, AIR's budget priorities, which are listed below, do not reflect this:

1. certification (which accounted for 53 percent of AIR's expenditures during fiscal year 1997)
2. continued airworthiness (35 percent)
3. rulemaking and policy development (12 percent)

Major Recommendation 1. It is critically important that the FAA and AIR conduct business in a new fashion with regard to aircraft certification and continued airworthiness. As an essential first step, AIR should revise its budget and manpower allocations to better reflect its mission priorities, which are as follows (FAA, 1998):

1. continued airworthiness and other activities related to continued operational safety
2. rulemaking and policy development
3. certification

Recommended Safety Management Process

The committee examined interrelationships between incidents and accidents and the current safety management process and identified areas that could be improved. Some of the basic elements of the current process either are not fully coordinated and integrated or are duplicated elsewhere. The committee, therefore, recommends an improved top level safety management process, which is illustrated in Figure 4-1 and described below.

Major Recommendation 2. It is essential that the FAA improve its safety management process. The FAA should work with the operators and manufacturers of large transport airplanes and engines to define and implement a proactive process that includes the following elements and tasks:

Key Elements
- data collection
- database management
- risk analysis
- risk management/action
- monitoring effectiveness

Specific Tasks
- Manufacturers, with the advice and consent of operators and the FAA, should define data requirements and processes for sharing data. Comprehensive FOQA (flight operations quality assurance) systems similar to BASIS (British Airways Safety Information System) should be used as a starting point.
- Operators should provide required data, as agreed upon.
- Manufacturers should solicit data from additional sources, such as the NTSB, ICAO, and National Aeronautics and Space Administration, to augment the operational database.
- Manufacturers, with oversight from the FAA and the assistance of operators, as required, should collect, organize, and analyze data to identify potential safety problems.
- Manufacturers should recommend corrective action for potential safety problems and seek consensus by operators. The FAA should make sure that actions proposed by manufacturers and operators will be effective, making regulatory changes and mandating compliance, as appropriate.
- Manufacturers and operators, with oversight from the FAA, should monitor the effectiveness and timeliness of corrective action and the safety management process (see Figure 4-1).

The thrust of this recommendation is that industry should collect, organize, and analyze safety data and take appropriate corrective action to protect the safety of the fleet. The FAA should not independently collect, organize, or analyze safety data for large transport aircraft. Instead, the FAA should oversee the entire process, providing direction, assessing the accuracy and objectivity of industry's risk analyses, and mandating corrective action, as appropriate. The overall objective is a more effective safety management process that routinely monitors operations and maintenance, uses data on incidents and other abnormalities to identify potential hazards proactively, and takes corrective action before those hazards cause an accident.

Many systems are currently used by industry and the FAA for generating and collecting data. Because most of these systems are not coordinated, however, there is a good deal of duplication, and much of the data cannot be used effectively. These systems consume scarce resources that could and should be applied to other safety efforts. The recommended

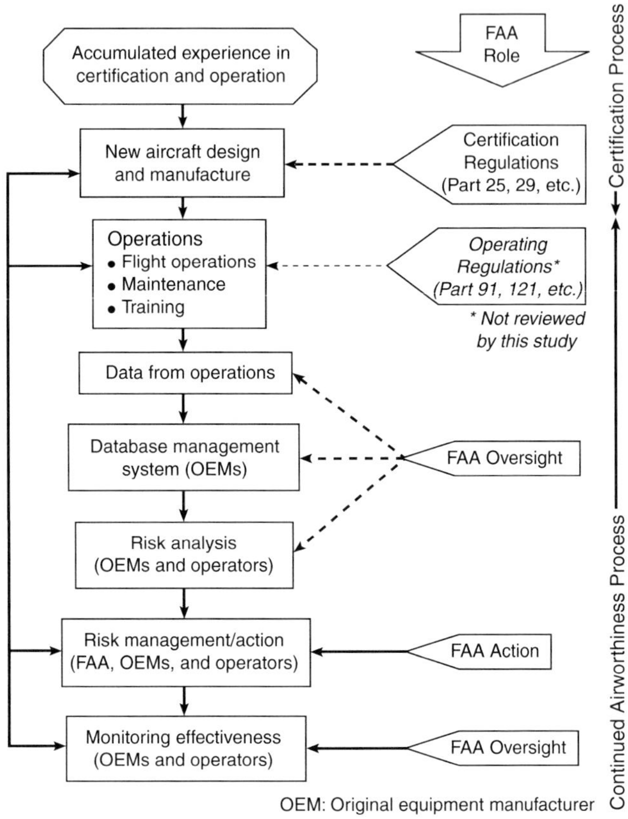

FIGURE 4-1 The recommended process for aircraft certification safety management.

safety management process would greatly reduce the number of systems and improve coordination among those that remain.

The value of some databases is limited because of poor data quality and difficulties in interpretation. For example, although accidents and incidents are caused by a chain of events involving many cause factors (see Chapter 3), most data collection and monitoring systems are not formulated to identify hazards that may arise from unusual combinations of factors that may not individually present a significant hazard.

Operators and manufacturers have much greater access to, knowledge of, and experience with their aircraft than the FAA. In addition, they may already have systems in place to collect the data needed for comprehensive safety management.

A missing element in the current safety management process is a widely accepted risk analysis system or methodology, which is necessary to allocate resources appropriately and to define timely and effective corrective action. The process recommended by the committee would correct this problem.

A credible and effective safety management process must accurately monitor, measure, and communicate the effectiveness of corrective action. This is especially important for overcoming industry doubts about proposed actions that would disrupt airline operations or reduce the profitability and competitive standing of operators or manufacturers. Accurate information on the cost and effectiveness of remedial action would also put the FAA in a better position to justify its own priorities and allocate more resources to activities with the most potential for improving aviation safety.

The recommended process could be implemented solely through regulatory changes that would require industry compliance. However, this approach would delay implementation for years because of the time it takes to make regulatory changes (see Chapter 6) and because of industry concerns that such changes would impose upon them an unproven system that would increase their costs and regulatory requirements. Therefore, the committee recommends that the process be implemented voluntarily, as much as practical. This would allow detailed procedures to be tested and improved while regulatory proposals are being formulated. A voluntary process would also help build the trust necessary for increased sharing of safety information, which will be essential to maximize the effectiveness of the recommended process.

In developing its recommendations, the committee considered several possible approaches to improving aviation safety. For example, the current safety management process, which has achieved an excellent safety record, could be continued with only minor changes. However, based on recent trends in the accident rate, this seems unlikely to achieve the desired safety improvements.

Another possibility would be to have a single organization collect and analyze safety data for all civil aircraft instead of sharing this responsibility among airlines and aircraft manufacturers, as recommended by the committee. Assigning this task to one organization, however, would separate the analysis function from the manufacturers, who have the most detailed insight into the design of their aircraft. In addition, establishing and maintaining a new organization to take on this massive task seems less practical than enhancing manufacturers' current capabilities. Manufacturers and operators already share a great deal of information derived from safety analyses, and comprehensive, widely accepted FOQA systems, such as BASIS, are increasing the flow of safety data and information. However, it seems unlikely that manufacturers would agree to share their proprietary design data, which an outside organization would need to conduct detailed safety analyses.

The committee also concluded that mandatory implementation of the recommended safety management process is a good long-term goal, but voluntary participation is essential for near-term impact.

Recommendation 4-1. In parallel with efforts to make appropriate regulatory changes, the FAA should expeditiously negotiate binding letters of agreement with manufacturers and operators to implement as much of the recommended safety management process as possible.

DATA COLLECTION

High quality data are essential to the effectiveness of the entire safety management process. Safety-related data are the foundation of the analytical processes used to prevent aircraft incidents or accidents. The type of data collected should be tailored to these analytical processes to ensure that enough of the right kind of data is collected and that the database management system is not overwhelmed with unnecessary data (see Appendix D).

Currently, data are generated and/or collected by many organizations (airlines, manufacturers, regulatory agencies, pilots, repair facilities, investigative agencies, independent agencies, and others). More than 80 large databases of aviation safety data are being used worldwide, some of which are mandated by regulatory authorities. Most of the data, however, are collected voluntarily by industry or government organizations because of their interest in aviation safety. For example, various FAA offices maintain a variety of aviation safety databases.

Maintaining the large number of current databases requires significant personnel and fiscal resources. Despite this investment, however, no existing database or collection of databases is fully satisfactory. In general, data collection efforts are fragmented, with individual efforts focused on different goals and objectives. For these reasons, current databases are unlikely to provide the comprehensive, high-quality data necessary to prevent incidents and accidents. For example, considerable data are collected by the FAA in the form of service difficulty reports from operators; pilot reports; and confidential safety reporting systems operated by the FAA and the National Aeronautics and Space Administration. However, much of this data is reported voluntarily or inconsistently, so a set of reports that seems to indicate that a particular aircraft system has developed a new problem may simply reflect a decision by one operator to start reporting a problem that has been present for some time. In this environment, developing a comprehensive and accurate understanding of system malfunctions is difficult. Yet such an understanding is essential for the development of a proactive safety management process that accurately predicts risks and identifies corrective action before an accident takes place.

Government and industry are now exploring ways to construct advanced databases, such as the Global Analysis and Information Network (GAIN), which would serve as a single global database. The development of GAIN has been hampered, however, by disagreements in the global community on which data should be collected, how the data should be standardized, how they should be shared and disseminated, and so on. At best, reaching consensus on these issues will probably take several years, and implementing the agreed-upon course of action will probably take several more years. Fortunately, other options exist for making significant near-term improvements in data collection.

The committee believes that the best source of most safety data is aircraft operators (including their maintenance organizations, whether they are part of the operators' organizations or outside contractors). Operational (and maintenance) experience generates the most important data for safety analyses. To be complete, however, these data should be supplemented with data from regulatory inspectors, accident and incident investigators, and manufacturers' design and test engineers. The data should be made available to manufacturers and the FAA, as appropriate.

The use of standardized formats for reporting data would facilitate data collection, database management, and the other elements of the safety management process. Thus, standardization is an appropriate topic for discussion between the FAA and industry. However, difficulties caused by nonstandardized data can be overcome, and implementation of the overall process should not be delayed by lack of consensus on standardized data formats.

The manufacturers of large transport airplanes and engines seem most likely to be able to manage comprehensive aviation safety databases. Unlike operators or regulatory authorities, manufacturers collect data globally, which gives them a larger set of data than national regulatory agencies or individual operators can collect. Also, manufacturers have the equipment to store and analyze these data and have the most detailed understanding of their own products.

Many individual aircraft manufacturers and operators are voluntarily installing QARs (quick access recorders) in new and existing aircraft. Unlike flight data recorders, QARs are usually not crashworthy. However, they can record up to 400 aircraft parameters and store data for many flight hours. The data from QARs are also easily accessible.

QARs are generally used to monitor the performance of aircraft and engine systems for operational and maintenance purposes. They can also be used to evaluate crew actions and performance. FOQA (flight operations quality assurance) programs use ground-based computers for routine analysis of operational data from QARs or digital flight data recorders (DFDRs). Like traditional flight data recorders, DFDRs provide data on aircraft and flight conditions for accident analyses, but DFDRs record much more information.

FOQA programs enhance flight safety by providing more information about, and greater insight into, the total flight operations environment through automated recording and analysis of flight data. In 1991, before any U.S. airlines had established FOQA programs, the FAA sponsored a study of FOQA programs used by foreign airlines. This study determined that "the appropriate use of FOQA data by airlines, pilot associations, and aircraft and equipment manufacturers would result in a significant improvement of flight safety by identifying operational irregularities that can foreshadow accidents and incidents" (FSF, 1992). The FAA subsequently established voluntary pilot programs with several U.S. airlines to document the safety and cost benefits of FOQA programs, assess technology alternatives, develop guidelines for FOQA programs in the United States, and address organizational strategies for the use, protection, and management of FOQA data and information derived from that data.

BASIS (the British Airways Safety Information System) is a comprehensive FOQA program. In addition to automated flight data, BASIS collects data from engineering reports, incident and accident reports, maintenance human factors reports, and flight crew human factors reports. The large pool of data submitted by participating airlines improves the capability of analyses to correlate data and identify rare phenomena that may have gone undetected in the past until they caused a serious incident or accident. Because the data collected by BASIS comes from many different types of aircraft, BASIS and systems like BASIS can also facilitate the exchange of lessons learned among manufacturers. This is important when a problem is relevant to aircraft produced by more than one manufacturer.

Analysis of data by BASIS considers cost and risk factors and produces targeted reports for flight crews, engineering organizations, maintenance organizations, regulators, and others. Although manufacturers are not directly involved in the analysis of data by BASIS, having access to a comprehensive data collection system like BASIS helps manufacturers improve the efficiency of their own data gathering efforts. Airbus and more than 100 airlines worldwide participate in BASIS to varying degrees, making it the largest such program currently in use. Boeing is also interested in obtaining data from BASIS.

In addition to BASIS and the pilot programs sponsored by the FAA, some airlines have implemented their own FOQA programs. The committee's recommendation to implement BASIS-like FOQA systems acknowledges BASIS's record as a widely accepted and comprehensive program. However, this recommendation is not an endorsement of BASIS over other similar systems, none of which the committee examined in detail.

Recommendation 4-2. As the recommended safety management process is implemented, the FAA should eliminate internal efforts to collect and store data for aircraft manufactured by companies with whom agreements have been reached in accordance with Recommendation 4-1. Resources currently used for those purposes should be redirected to AIR's other safety-related functions.

DATABASE MANAGEMENT

Database management systems (DBMSs) are computer-based systems used to store, manage, retrieve, and update data that is stored in a database. DBMSs ensure the integrity of databases while allowing simultaneous access by many users. Relational, object, and object-relational databases are described below.

Relational Database Management Systems

Relational DBMSs, which are the most widely used, are available from many vendors. In a relational DBMS, information is stored in a series of simple tables, much like a series of spreadsheets. The rows in each table represent objects, such as airplanes, airports, or people. The columns in each table describe one facet of the object, for example, name, serial number, or number of engines. A relational database is a collection of one or more of these tables.

Relational DBMSs are relatively simple to build and are very good for traditional business applications. Standard Query Language (a computer language) is widely used for accessing information in relational DBMSs. These systems work well with simple types of data and predefined operations and queries. However relational DBMSs are limited and difficult to work with when the information in the tables is interrelated and when stored data includes video images, documents, pictures, and other complex data.

Object Database Management Systems

Object DBMSs take a real-world approach to the definition and storage of data. These systems emulate the environments in which stored objects exist. For example, objects that could be part of an object DBMS for accident investigations include accidents, incidents, airplanes, and crews. Within the system, it is very easy to create and maintain connections between related objects, and most object DBMSs can accommodate complex data types, such as video images, pictures, and documents.

Although object DBMSs are very good for storing complex objects, they are more difficult to use than relational DBMSs because they usually cannot be accessed through a simple query language. In addition, these systems are not well suited to environments that have to support large numbers of users and large numbers of queries.

Object-Relational Database Management Systems

Relational and object DBMSs both have strengths and weaknesses. Object-relational DBMSs attempt to combine the best features of both. Object-relational systems can support the definition and maintenance of complex objects and complex data types, and they can provide easy access to the information using a variation of the Standard Query Language. An object–relational DBMS also has fewer performance problems than object DBMSs because it can support both user and system-defined indexes to speed up transaction processing. The object-relational DBMS is the newest of the three systems described here, and only a limited number of vendors offer systems of this type.

Choosing the Right Database Model

It is easy to build a database that does nothing to enhance an organization. Some organizations seem to believe that simply putting information on line will solve information problems. Unless users have a clear understanding of the objectives, however, electronic information is destined to sit on a computer—unseen and unused.

DBMSs are tools. The first step in choosing a DBMS is to identify the most serious problems, determine what data are needed to address these problems, and determine what information will be stored. The next step is to understand the nature and capabilities of the people who will build, use, and maintain the system. Only then is it appropriate to select a DBMS. For instance, aviation safety analysis includes evaluations of a large number of specific events in various combinations and permutations to determine the conditional probability of a serious accident. Thus, a safety DBMS should have the capability to store and manipulate complex objects and data types efficiently and effectively. This is a difficult challenge.

Recommendation 4-3. Manufacturers should establish aviation safety DBMSs using the state-of-the-art data management technologies that are best suited to continued airworthiness applications. The most suitable type of DBMS currently available is the object-relational DBMS.

RISK ANALYSIS

An effective safety management process should include risk analysis to provide a sound basis for risk management, which involves making decisions and taking action to reduce risks and, in the context of this study, to improve safety. Risk analysis involves three steps:

1. listing possible outcomes (favorable and otherwise)
2. estimating the consequences associated with each outcome
3. estimating the probabilities of each outcome

In this discussion, a *consequence* is defined as a numerical measure of the loss or harm associated with an adverse outcome. The scale chosen to measure loss or harm must allow for meaningful addition and multiplication over different events. Examples of appropriate scales are costs in dollars, losses in productivity, and reduced life expectancy. A more universal scale is utility (or disutility), which is a unitless parameter with a value between 0 and 1.

The term *risk* means the probability that a particular adverse event (or outcome) will occur during a stated period of time or will result from a particular challenge (The Royal Society, 1992). As a probability, risk must obey all the formal rules of combining probabilities and is also subject to the vagaries of interpretation. (See Appendix E for a brief discussion of these interpretations.) Because the probability that systems will perform as expected is closely linked to their reliability and maintainability, these concepts lie at the heart of risk analysis and risk management. Failure data analysis is also important—in order to correctly interpret data on system performance.

Paradigm for Risk Analysis and Risk Management

Risk analysis and risk management are closely linked, as illustrated in Figure 4-2. In many cases, risk management decisions are based solely on the personal judgment and expertise of responsible personnel. In other cases, the complexity of the issues and the magnitude of the potential consequences may warrant a more rigorous approach. In these situations, the decision maker may commission a risk analysis before selecting a particular course of action.

Referring to Figure 4-2, suppose a management decision is needed to resolve a safety-related issue. In this case, management must select one of two possible decisions, D_1 or D_2. Decision D_1 leads to one of three outcomes, O_1, O_2, or O_3, whereas D_2 leads to the outcomes O_1, O_3, or O_4. The product of the probability of occurrence of each outcome and the severity of the consequences of the outcome is known as the *expected utility* of the outcome. The best method for maximizing total expected utility involves the use of decision trees to calculate consequences.

As a hypothetical example, D_1 might be a decision to install a smoke detector in the hold of all cargo planes, and D_2 might be a decision not to install smoke detectors. At a top level, the possible outcomes include the following:

- O_1—no fire in the hold
- O_2—fire in the hold and the smoke detector functions properly
- O_3—fire in the hold and the smoke detector fails
- O_4—no inflammable material in the hold

After the outcomes have been identified, risk analysis is used to evaluate the consequences of each outcome by calculating its utility or cost. $U(D_1, O_3)$ would be the consequence of installing a detector that fails when there is a fire in the hold. In this case, the consequences of decision D_1 would include the costs of acquiring, maintaining, and operating the smoke detectors, as well as the costs generated by a fire that is not quickly detected.

Once the outcomes and their consequences have been defined, the next step is to determine the probability (P) that each outcome will occur. $P(D_1, O_2)$ is the probability that outcome O_2, a fire in the hold of an aircraft with a properly functioning smoke detector, will occur after a decision is

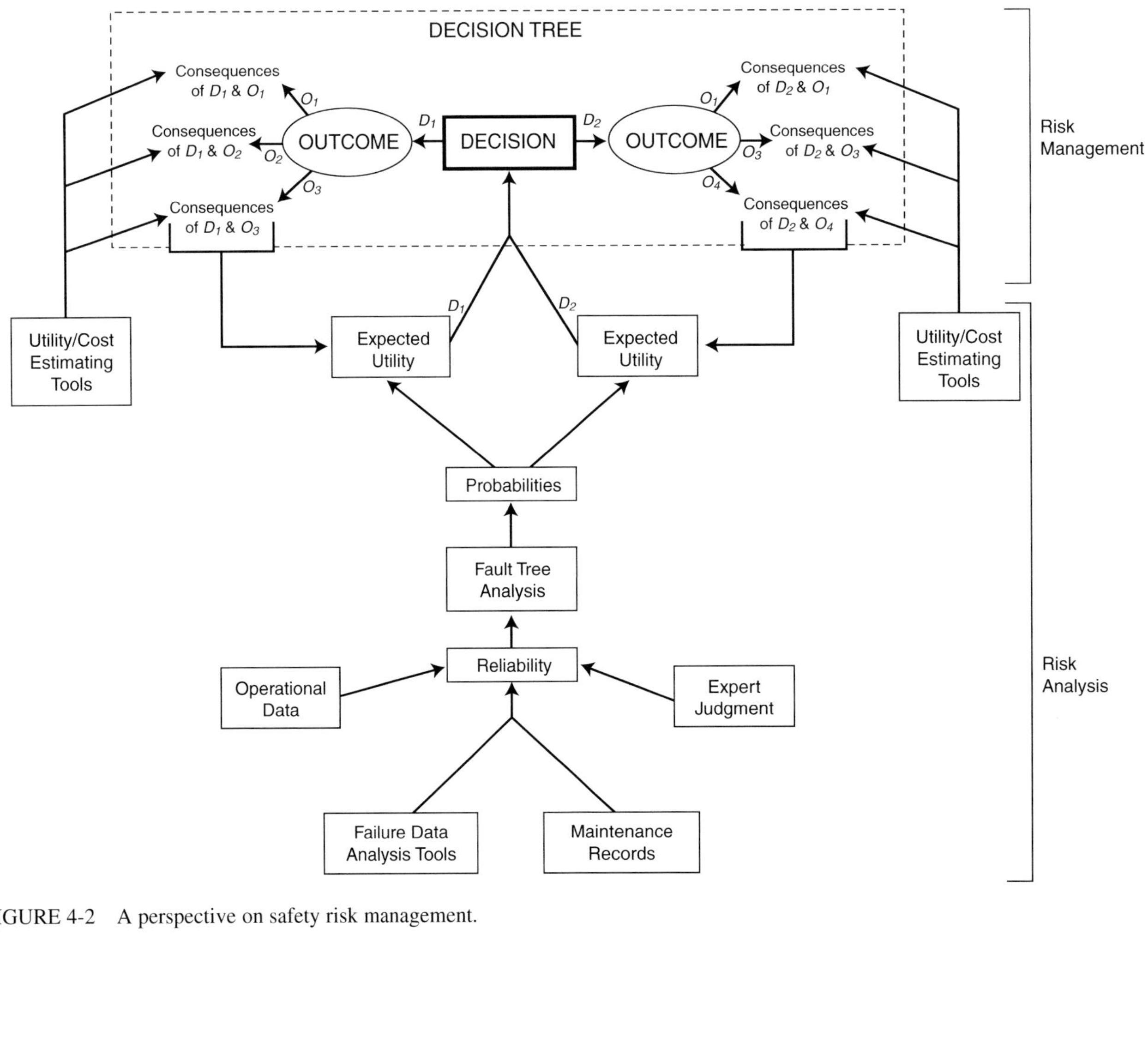

FIGURE 4-2 A perspective on safety risk management.

made to install a smoke detector (D_1). The probabilities are obtained via fault trees (see Appendix E) and reliability data, which are themselves based on failure data analysis tools, expert judgment, operational data, and maintenance records. Once the probabilities have been determined, expected utilities $E[U(D_1)]$ and $E[U(D_2)]$ can be computed. For example, $E[U(D_1)]$ would be the sum of the expected utilities for each outcome associated with decision D_1.

Based on this analysis, an informed risk management decision can be made. For the case shown in Figure 4-2, decision makers would logically choose decision D_1 if the expected utility $E[U(D_1)]$ is greater than the expected utility of $E[U(D_2)]$; otherwise they would choose D_2. If $E[U(D_1)] = E[U(D_2)]$, then either decision could be chosen.

Risk analyses are based on quantitative procedures. However, because of the difficulty of accurately assessing probabilities and expected utilities, risk assessments today are imprecise and require some subjective inputs. Risk analyses consider how combinations of unusual circumstances could impact expected outcomes and consequences, but for a given problem, it may be difficult to define an appropriate set of abnormal conditions. In addition, judging the severity of the consequences for some outcomes may be subjective, and it may be difficult to obtain a consensus among the FAA, industry, and the general public about the severity of consequences or how to compare consequences. For example, professional pilots may believe that airlines have overestimated the costs of taking corrective action, which would lower its estimated cost-effectiveness, or decision makers and the public at large may not agree on the relative importance of expected utilities.

Risk Analysis and the Air Transport Industry

The committee met with aircraft manufacturers (Boeing and Airbus), airline operators (Alaska Airlines and Delta Airlines), and engine manufacturers (Pratt and Whitney and General Electric) to discuss industry approaches to risk analysis and risk management. Some committee members also have extensive industry experience. The committee concluded that industry generally does a good job of using risk analyses to identify and understand risks, although different companies use different methods. For example, manufacturers pay careful attention to engineering and operational details that affect safety and reliability. Fault tree analysis—which is widely used for reliability analyses in many other industries—was conceived in 1961 at Bell Laboratories and refined by Boeing so that the quantitative portions could be done by computer (Roland and Moriarty, 1983).

Ongoing efforts by the aviation industry to improve the safety of existing and new aircraft include many quantitative techniques. The committee was unable to determine, however, the extent to which manufacturers use dependency models in their reliability assessments. The committee did not assess the effectiveness of industry processes either for demonstrating software reliability, which is a growing challenge, or for integrating individual risk analyses into a comprehensive package for risk management.

The BASIS system mentioned earlier groups expected utilities into blocks, places the blocks into one of several categories, and proposes actions for each category. This approach comes closest to implementing the paradigm for risk analysis and risk management described at the beginning of this section. BASIS also attempts to describe the human factors aspects of airline operations using fault-tree-like approaches.

Risk analysis and risk management activities by major engine manufacturers seem to be focused on engine reliability, which is extremely high. The Continued Airworthiness Assessment Methodologies (CAAM) is an engine reliability and failure data analysis tool used to identify and prioritize unsafe conditions. CAAM is reactive in the sense that it depends on data from incidents and other reported problems, and it cannot react to situations for which operational data are not available. More importantly, however, CAAM is proactive in the sense that it uses data from minor abnormalities to predict more serious problems. In other words, systems like CAAM that are reactive to incidents may be proactive to accidents.

Overall, the committee believes that industry's risk analysis efforts, including BASIS, other systems similar to BASIS, and CAAM, have been laudable. Safety boards in individual companies formalize the safety management process, add discipline, and generate safety decisions in close cooperation with the FAA. Aircraft manufacturers have established a detailed and generally thorough process for quantifying risk, with a strong emphasis on collecting large volumes of data and heavy reliance on fault tree analysis and failure data analysis. Additional use of advanced analytical tools may improve the effectiveness of industry's analyses. For example, the committee saw little evidence that expert opinion or scientific judgments are formally incorporated into reliability assessments, and it may be possible to improve failure data analysis by relying more on scientific and engineering information to supplement operational and maintenance data.

Risk Analysis and the Federal Aviation Administration

Risk analysis and risk management are important tools for understanding risks, defining acceptable levels of risks, and reducing risks. Establishing consensus about the purpose, role, and capabilities of risk analysis is an important prerequisite for an effective risk analysis program. For example, some FAA officials believe that, because of its subjective nature, risk analysis should not be used to determine if a condition is unsafe and warrants mandatory action. These officials would use risk analysis only to establish a time frame for correcting an unsafe condition—not to determine if the condition itself is unsafe. The committee, however, believes that proactively minimizing and avoiding the risk of serious incidents or accidents requires analysis tools, such as macrolevel fault trees, that integrate information from manufacturers and operators.

Suggestions have been made both inside and outside the FAA that the FAA should adopt a more formal approach to risk management, and the FAA is currently improving its risk analysis capabilities. The committee believes that the FAA should focus its efforts on improving its ability to oversee industry's risk analysis, while encouraging industry to continue developing its capabilities.

RISK MANAGEMENT/ACTION

Risk management involves choosing the best combination of advantages and disadvantages from several alternatives in the presence of uncertainty. This is analogous to cost-benefit analysis in the presence of uncertain outcomes. In short, risk management boils down to choosing the best *available* alternative, without regard to alternatives that are not available (Derby and Keeny, 1981).

The *risk management/action* function of the recommended safety management process is directly dependent on the *risk analysis* function for determining probabilities and costs. Risk management actions can be initiated by the FAA through modified regulations, new regulations, and ADs; by manufacturers through service bulletins; and by operators through operator-initiated engineering actions. AIR can initiate regulatory actions related to modifications of existing aircraft, future aircraft designs, and/or requirements for aircraft maintenance programs. The committee believes that manufacturers should use the results of risk analyses to recommend corrective action and seek consensus by operators. The FAA should make sure that actions proposed by manufacturers and operators will be effective and mandate compliance, as appropriate. The following discussion summarizes the current environment relative to ADs, service bulletins, and rulemaking and offers specific recommendations for improvement.

Airworthiness Directives

The AD review process is frequently and simultaneously criticized by external parties with diametrically opposed views. The media, Congress, NTSB, and other government officials often criticize the process for being too time consuming. Airlines and manufacturers often complain that the time allotted for logistical, engineering, scheduling, and maintenance actions is too short.

The FAA is compelled by law and regulation to provide public notice of proposed actions, generally by publication in the Federal Register. Unless immediate action is needed to address an urgent safety issue, the FAA must provide sufficient time for public comment on the appropriateness and effects of proposed ADs and other rulemaking actions, and the FAA must complete a written review and analysis of public comments in terms of effectiveness, cost, and time. The existing process has been criticized for not being more thorough in responding to public comments, for limiting the degree to which alternative actions are considered, and for not being more accurate in estimating the time and cost for accomplishing proposed actions. Current procedures subject proposed regulatory actions to peer review, but the committee believes this peer review process should sometimes be more thorough. In addition, the accuracy of the FAA's time and cost estimates could be improved by developing models consistent with industry data and experience.[1]

Recommendation 4-4. Consistent with regulatory procedures, the FAA should develop a more accurate methodology for assessing the costs and benefits of potential ADs and other rulemaking actions, as appropriate. In particular, the FAA should work with industry to develop more realistic and more reliable models for estimating time and cost.

Manufacturers' Service Bulletins and Regulatory Actions by Foreign Airworthiness Authorities

Manufacturer's service bulletins include procedures, lists of materials, and specifications for technical modifications or inspections to aircraft and aircraft systems. Service bulletins are developed by aircraft manufacturers and equipment manufacturers for use by airlines, repair stations, and other organizations authorized by the FAA to make the modification. Service bulletins are developed by manufacturers to improve aircraft characteristics in terms of safety, reliability, operating costs, etc. Often, the recommended timing and method of implementing a service bulletin are established with inputs from operators of the affected equipment based on their relevant service experience.

If the FAA determines that implementation of a safety-related service bulletin should be mandatory, it generally publishes an AD that defines implementation of the service bulletin as the means of compliance. The FAA generally allows operators to request approval to comply with ADs by alternate means. To be approved, these requests must demonstrate equivalent levels of safety, integrity, and airworthiness. Industry requests for alternate means of compliance are most commonly generated for the following reasons:

- The operator may want to use equivalent materials that are not specified in a particular service bulletin and/or AD. Such a request could be driven by economic concerns or by the lack of availability of the specific materials delineated in the service bulletin.

[1]Rulemaking processes and problems are also discussed in Chapter 2, Chapter 6, and later in this chapter.

- The operator defines an equivalent engineering fix or alternate means of compliance that provides an economic, maintenance, or timing advantage over the action specified by the AD.

All operators are required to maintain records that define the configuration of their aircraft relative to regulatory, manufacturer, and operator-initiated engineering actions. However, operators are not required to use a standard format for verifying the configuration of individual aircraft. Consequently, an extensive records search is sometimes required to determine the actual configuration, particularly for aircraft that have had multiple changes in ownership or changes in ownership between foreign and domestic operators. Defining the configuration of an aircraft may be particularly difficult if the aircraft is not available for inspection (e.g., when it has been destroyed in an accident). In this case, determining the aircraft configuration could be essential to determining corrective action that would prevent future accidents.

When the FAA initiates an AD applicable to aircraft that were originally certificated by the FAA (i.e., U.S.-manufactured aircraft), the regulatory agencies of many other nations adopt the AD immediately. The reverse is not true, however. Because of legislative, administrative, and regulatory requirements, the FAA must provide the same level of internal and public review for airworthiness actions issued by other nations as it does for its own regulatory actions. These requirements have helped create a backlog of hundreds of regulatory initiatives within the FAA. This situation generates a dilemma for U.S. operators of aircraft that were originally certificated by other nations (e.g., a U.S. airline operating Airbus aircraft). If U.S. operators do not implement the airworthiness action as specified by the foreign regulatory agency, they could be operating their aircraft with a lower level of airworthiness than aircraft operated by foreign operators. However, if U.S. operators implement the airworthiness action as specified, the FAA may later approve other compliance requirements that would invalidate their actions.

Recommendation 4-5. To eliminate the regulatory backlog and the ambiguities about implementing airworthiness actions of foreign regulatory authorities, the FAA should expeditiously determine what regulatory action, if any, it will propose in response to foreign airworthiness actions. The FAA should initiate its regulatory response no later than two weeks after receiving notice of a foreign airworthiness action.

Rulemaking

The FAA can react immediately to critical safety and security problems by issuing "immediately adopted" rules. The normal process for issuing a new regulation or modifying an existing regulation, however, is laborious and time consuming—quite often taking 5 to 10 years from start to finish. This is caused partly by the limited availability of personnel for rulemaking activities, the large volume of pending actions, and the procedural requirements imposed by the executive and legislative branches of the government (see Chapter 2). Several years ago, in an attempt to reduce delays, the FAA established the ARAC (Aviation Rulemaking Advisory Committee). The purpose of the ARAC is to allow FAA staff, industry experts, and other interested parties to reach an early and fully informed consensus on the need for and content of proposed rulemaking actions. However, this approach has not yielded the intended time savings, and long delays in the rulemaking process remain a critical barrier to improving the safety management process (see Chapters 6).

Corrective Action vs. Blame

After an incident or accident occurs, cause factors should be identified and effective corrective action should be implemented in a timely fashion. The process of assigning blame, on the other hand, often does not reduce the risk of future incidents and accidents. It is easy, but unprofitable, to associate "causes" with "blame." According to John K. Lauber, former member of the NTSB, current efforts to improve aviation safety are hampered by

> . . . a blurring of the distinction between incident and accident causes on the one hand, and legal, economic, and moral responsibility on the other. In our culture, we seem to be unable to deal with problems of any importance without assessing blame, and perversely will happily march over the cliff if we are certain that we know who to blame for our imminent demise. All too frequently, our search for someone to blame takes real priority over our search for solutions, and this seems especially true in matters of aviation safety (Lauber, 1989).

The committee believes that the safety management process recommended in this report would help shift the focus of incident and accident investigations away from the question of who is to blame. Instead, the process would focus on identifying corrective action to prevent similar problems *and* their consequences.

MONITORING EFFECTIVENESS

One of the critical elements of any effective control system is a feedback loop. The feedback loop measures the effects of system variables on the outcome of the system and indicates how the system can be improved. Feedback enables the system operator to maximize the performance of the system and evaluate the effectiveness of previous attempts to improve system performance. Considerable feedback data relevant to continuing airworthiness are available to the FAA. The committee believes that evaluating the effectiveness of ADs and other required actions would be greatly facilitated if industry, with FAA oversight, placed a higher priority on monitoring the effectiveness of corrective action.

Changes in accident rates can provide a sense of the overall effectiveness of ADs but cannot be used to measure the effectiveness of individual ADs. However, it is relatively easy to assess the effectiveness of most ADs. For example, if a safety-related part repeatedly fails prior to regularly scheduled inspections, one or more ADs may be issued to increase the inspection frequency and, later, to install an improved part after one has been designed, tested, and approved for use. The effectiveness of such ADs can be verified by continuing to track unexpected failures.

By monitoring the effectiveness of ADs, the FAA could enhance risk reduction by placing its highest priority on areas that are shown to be most effective. Industry could facilitate the FAA's efforts to monitor the effectiveness of airworthiness actions by agreeing upon and maintaining a standardized summary of aircraft configuration that records the implementation of voluntary service bulletins and airworthiness actions issued by foreign regulatory authorities, as well as ADs.

An important aspect of the recommended safety management process is a comprehensive FOQA system, such as BASIS, for collecting data from automated flight data recorders, maintenance reports, incident and accident reports, etc.

APPROVED DESIGN ORGANIZATIONS

With few exceptions, existing legislation and regulations do not require applicants for type certificates, amended type certificates, STCs (supplemental type certificates), TSOAs (technical standard order authorizations), PMAs (parts manufacturing approvals), and other approvals to show that they have the technical qualifications to develop a safe design or to conduct the engineering evaluations and certification tests necessary to show compliance with applicable FAA airworthiness standards. In addition, applicants for type certificates and STCs are not required to establish or maintain technical organizations to monitor, evaluate, and propose corrective action in response to operator reports of safety problems for which they are responsible. Of course, major aircraft manufacturers have skilled and experienced engineering organizations, and they do not need regulatory encouragement to closely monitor the safety performance of their products. However, the lack of statutory authority and implementing regulations has two negative effects: it limits the FAA's ability to take advantage of the capabilities of certificate holders' design organizations, and it requires the FAA to spend considerable resources each year on "false starts" by applicants that do not have the technical qualifications to complete the application process. The latter situation arises most frequently with applicants for STCs and PMAs.

The process by which the FAA regulates the production of aircraft, engines, and propellers is a model that could be applied to other aspects of the type certification process. Before granting a production certificate, the FAA evaluates the applicant's production quality control system. After a production certificate is issued, the FAA conducts periodic audits to make sure products are being manufactured in accordance with the approved quality control system. However, the FAA does not routinely make detailed inspections to determine if individual aircraft, engines, or propellers conform to the approved design and are eligible for airworthiness certification. In other words, the FAA promotes the safety of individual products by verifying that a safe and effective production system—that includes its own internal checks—has been established and is being maintained.

A similar approach should be used to promote the safety of product designs. The FAA should assess and approve the capabilities and procedures of an applicant's design organization rather than the current process, which requires FAA engineers to analyze independently the safety implications of new and modified designs. The design organizations of aeronautical engineering consulting firms, airlines, repair stations, and other organizations could also be approved by the FAA to enable applicants that do not have their own qualified engineering organizations to apply for STCs, TSOAs, and PMAs.

When the current type certificate process was developed, aircraft and engines were much smaller and less complex than they are today. In those days, it was feasible for the FAA to verify independently safety-related aspects of manufacturers' designs. Today, however, a major airframe manufacturer may employ as many as 8,000 engineers, flight test pilots, and inspectors to design, develop, and certificate a new wide-body passenger jet. These large staffs are necessary to investigate the design complexities of modern aircraft. The number of labor hours invested by a manufacturer in designing a large new jet may be several hundred times greater than the number of labor hours the FAA has available to verify the safety of the aircraft design. This huge discrepancy raises a question about the FAA's ability to analyze independently new aircraft designs and locate safety-related design flaws that are subtle enough to have escaped the attention of the manufacturer's much larger design team. In fact, as designs have become more and more complex, the FAA has had to rely more and more on spot checks of new designs (instead of comprehensive reviews).

The committee believes that design safety would be enhanced if the FAA devoted its engineering resources to promoting the safety and efficacy of manufacturer's design teams and processes, rather than trying to identify problems in specific designs. The FAA should examine the technical qualifications and integrity of design organizations, including their understanding of regulations and policies and their ability to properly implement them. Qualified organizations should then be certificated as approved design organizations (ADOs), allowing them to make detailed findings of compliance in accordance with published policies. FAA audits would verify continued compliance, in part by ensuring that

ADOs' level of involvement in specific projects is appropriate in light of the technical issues involved. Each ADO would be rated with limitations consistent with its technical capabilities and needs. For example, an applicant for an ADO to approve the designs of small interior parts such as tray tables or galley drawers would only need to demonstrate that it has the technical capabilities to determine that these products meet applicable airworthiness standards, and its authority would be limited accordingly.

Establishing a system of ADOs would expand the current system under which the FAA already delegates specified certification functions to *individual* designees, such as designated engineering representatives, as described in Chapter 2. In addition, FAR Part 21 already authorizes the FAA to designate qualified *companies* to perform selected type certification functions. However, except in cases where special exemptions are sought and approved, current regulations prohibit extending this authorization to manufacturers of transport category aircraft, turbojet engines with more than 1,000 pounds of thrust, propeller engines with more than 500 brake horsepower, and propellers for these engines.

Recommendations for removing this prohibition date back to 1966 but have not yet been implemented (FAA, 1966). A working group of the ARAC (Aviation Rulemaking Advisory Committee) is currently developing a draft rule to establish an "Organization Designated Authorization," which would extend the current delegation authorizations to manufacturers of large aircraft, large engines, and propellers for large aircraft (Federal Register, 1993). Adopting such a rule would improve the efficiency of the current certification processes. A more comprehensive restructuring of the process, which would include establishing ADOs, requires legislative authorization.[2]

AIR is also supporting nonregulatory approaches to improving the certification process by trying to define modified processes to achieve the following goals:

- early definition of applicable airworthiness standards, including special requirements for novel and unusual design features and exemptions where safety would not be compromised
- early agreement on what constitutes acceptable means of compliance and on findings of equivalent safety
- early completion of basic safety assessments to identify areas that require more detailed FAA involvement
- early agreement by the FAA and the applicant on a plan for completing the application process

[2]Chapter 447 of Title 49 of the U.S. Code would have to be amended to authorize ADOs. Appendix F contains the draft of a sample legislative amendment that could make this change.

Major Recommendation 3. AIR should promote aircraft safety by certifying the competency of applicants' design organizations rather than relying on the FAA's ability to detect design deficiencies through spot checks. The FAA should work with industry and Congress to obtain legislative and regulatory authority in a timely fashion to do the following:

- Certificate and rate ADOs and invest them with the responsibility for ensuring that applications for type certificates, type certificate amendments, STCs, TSOAs (technical standard order authorizations), and PMAs (parts manufacturer approvals) comply with applicable airworthiness standards. ADOs would be required to have the technical capabilities necessary for competently approving designs only within the limitations of their rating.
- Require ADOs and holders of production certificates to collect and analyze relevant safety data received from operators and to define corrective action in the event unsafe conditions are detected.
- Require applicants for design approvals to either hold an ADO certificate or employ the services of an ADO.
- As an interim step, give higher priority to the ongoing rulemaking action that would increase organizational delegation to manufacturers of large aircraft and engines under the FAA's current legislative authority. The FAA already uses this authority to grant organizational delegation to manufacturers of small aircraft and engines.

REFERENCES

Derby, S. L., and R. Keeney. 1981. How safe is safe enough. Risk Analysis 1(3): 217–224.

FAA (Federal Aviation Administration). 1966. Airworthiness Standards Evaluation Committee Report to the Administrator. Washington, D.C.: FAA.

FAA. 1998. Aircraft Certification Mission Statement. Federal Aviation Administration. Online. Aircraft Certification Service (AIR) Headquarters Office Home Page. Available: http://www.faa.gov/avr/air/hq/mission.htm. February 10, 1998.

Federal Register. 1993. Aviation Rulemaking Advisory Committee. Delegation System Working Group Tasking. 58: 1673.

FSF (Flight Safety Foundation). 1992. Flight Operations Quality Assurance Programs. Washington, D.C.: FSF.

Lauber, J.K. 1989. Human performance and aviation safety: some issues and some solutions. Accident Prevention Bulletin 46(4): 10–13.

Roland, H., and B. Moriarty. 1983. System Safety Engineering and Management. New York: John Wiley & Sons.

The Royal Society. 1992. Risk Analysis, Perception and Management. Report of a Royal Society Study Group. London: The Royal Society.

5

Human Factors

INTRODUCTION

Human factors issues, specifically human errors, contribute to more aircraft incidents and accidents than any other single factor. Human errors include errors by the flight crew, maintenance personnel, air traffic controllers, and others who have a direct impact on flight safety.

What lies behind human error is very frequently inaccurate situational awareness: the failure (for whatever reason) to evaluate an operational or maintenance situation properly. Thus, whenever the term human error appears, the reader should keep in mind that situational awareness, or the lack thereof, is usually the dominant factor. This can be a critical problem. As noted in Chapter 2, lack of situational awareness is a key factor in CFIT (controlled flight into terrain) accidents, which are responsible for more fatalities than any other type of aircraft accident.

This chapter discusses the relationships between human factors, environmental factors, and equipment factors in accidents and incidents; reviews current initiatives to reduce accidents and incidents associated with human errors or misunderstanding; and recommends steps the FAA can take to improve the effectiveness of its human factors work.

RELATIONSHIPS BETWEEN HUMAN FACTORS, ENVIRONMENTAL FACTORS, AND EQUIPMENT FACTORS IN ACCIDENTS AND INCIDENTS

Human factors are significant contributors in approximately 70 percent of all accidents and incidents. In a review of several databases, the committee found values in the range of 60 percent to 85 percent. These differences do not reflect on the integrity of the databases; they reflect the databases' different purposes and the understandable difficulties that arise from the substantial overlap of environmental, equipment, and human factors issues. This overlap, which is illustrated in Figure 5-1, is intrinsic to a complex system with a large number of possible accident and incident sources (primary and contributory). For example, adverse weather (or the threat of adverse weather) can contribute to an accident in many different ways. Weather information is generally, but not always, accurate; weather information provided to flight crews at dispatch and in flight is generally, but not always, timely; flight crew decisions based on available information are generally, but not always, made in accordance with prescribed procedures. There is no clear way, and indeed no practical need, to separate entirely environmental from operational factors.

Inaccurate situational awareness by the flight crew can arise in several different ways. Some examples are listed below:

- The flight crew may not have critical data necessary to adequately define its situation, which may lead to inappropriate decisions and, ultimately, an accident.
- The flight crew may have the data it needs but misinterpret the data.
- The flight crew may have the data it needs, properly interpret the data, and accurately define the situation, but it may not have the training, skills, or procedures to make proper decisions or to carry them out in the time available.

Automated features of flight control systems can improve situational awareness by reducing crew workload. However, automated actions that compensate for unusual flight conditions or equipment malfunctions can reduce situational awareness if the automated system masks the presence of abnormalities or does not clearly indicate what actions it is taking in response.

Aircraft must be designed so that, for all situations the flight crew can reasonably be expected to encounter, it will have the data it needs in an easily recognizable form that facilitates proper decision making. Furthermore, the aircraft should be designed to help the flight crew carry out necessary tasks, especially in emergencies when things are not as

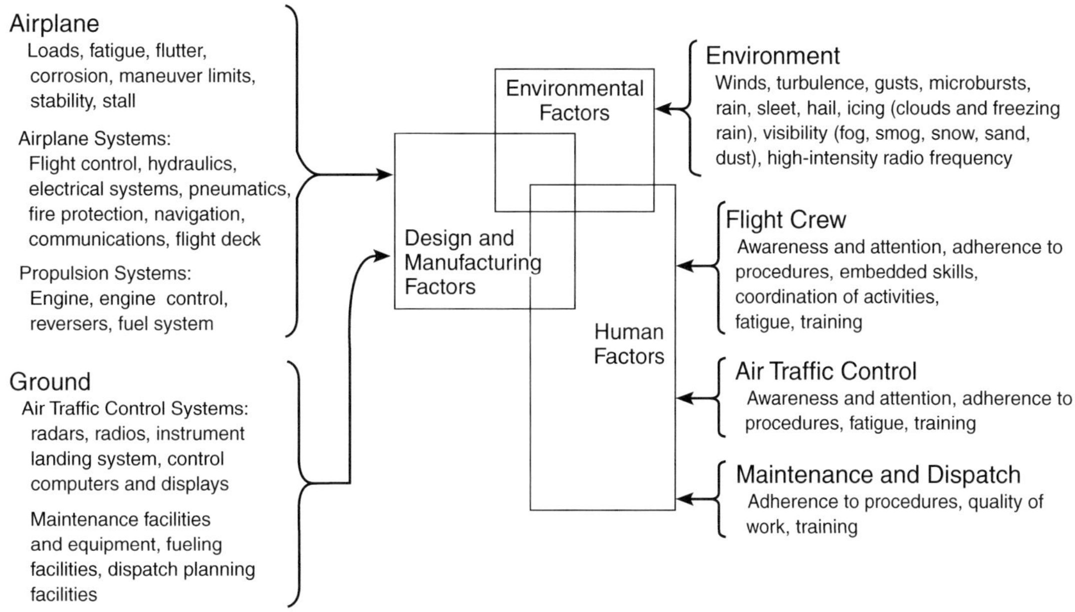

FIGURE 5-1 Elements for consideration in safety evaluations.

expected and safety depends on quick and correct actions by the flight crew. Except for the time pressure typically associated with in-flight emergencies, the same considerations apply to the actions of maintenance personnel.

CURRENT INITIATIVES TO REDUCE ACCIDENTS AND INCIDENTS ASSOCIATED WITH HUMAN ERROR

Many aspects of human factors are associated with the operational safety of commercial airplanes, including the following:

- design factors associated with aircraft controls, aircraft system controls, warning systems, air traffic control systems, flight deck, passenger seating and egress, etc.
- operational factors associated with the selection and training of flight crews, crew assignment policies related to the distribution of experienced personnel and the minimization of flight crew fatigue, checks on crew members' health, and policies on preflight information
- maintenance factors related to training maintenance workers; the clarity of maintenance procedures; and designing aircraft equipment and maintenance tools to make it easier for workers to perform maintenance, avoid errors, and detect abnormal conditions
- national and international regulatory factors associated with airworthiness standards, separation standards, and communications standards

Current processes, which are both thorough and complex, have resulted from a large accumulation of flight experience, analytical and computer studies, and reviews of human factors. All of this information represents a complicated web of interrelated factors that makes it difficult to define a clear and simple road map for progress. Complexity, however, is inherent in many human factors issues.

Figure 5-2 provides a greatly simplified view of human factors initiatives related to aviation. A much more detailed picture of the breadth and depth of current work and what needs to be done is available in the following publications:

- The detailed report of the FAA Human Factors Team, *Interfaces Between Flight Crews and Modern Flight Deck Systems* (1996), includes more than 50 well formulated recommendations.
- The *Proceedings of the FAA Workshop on Flight Crew Accident and Incident Human Factors* (1995) explores three human factors objectives.
- The April 1997 International Symposium on Aviation Psychology includes more than 300 papers on human factors associated with flight safety.
- The National Aeronautics and Space Administration report, *Principles and Guidelines for Duty and Rest Scheduling in Commercial Aviation* (1996), defines numerous general principles, specific principles, guidelines, and strategies for improving duty and rest scheduling practices.
- The NTSB review, *Flight-Crew-Involved Major Accidents of U.S. Air Carriers, 1978 through 1990* (1992), includes five broad recommendations.
- *Human Factors Guide for Aviation Maintenance* (FAA, 1997), published by the FAA's Office of Aviation Medicine, presents basic concepts on reducing human errors in maintenance.

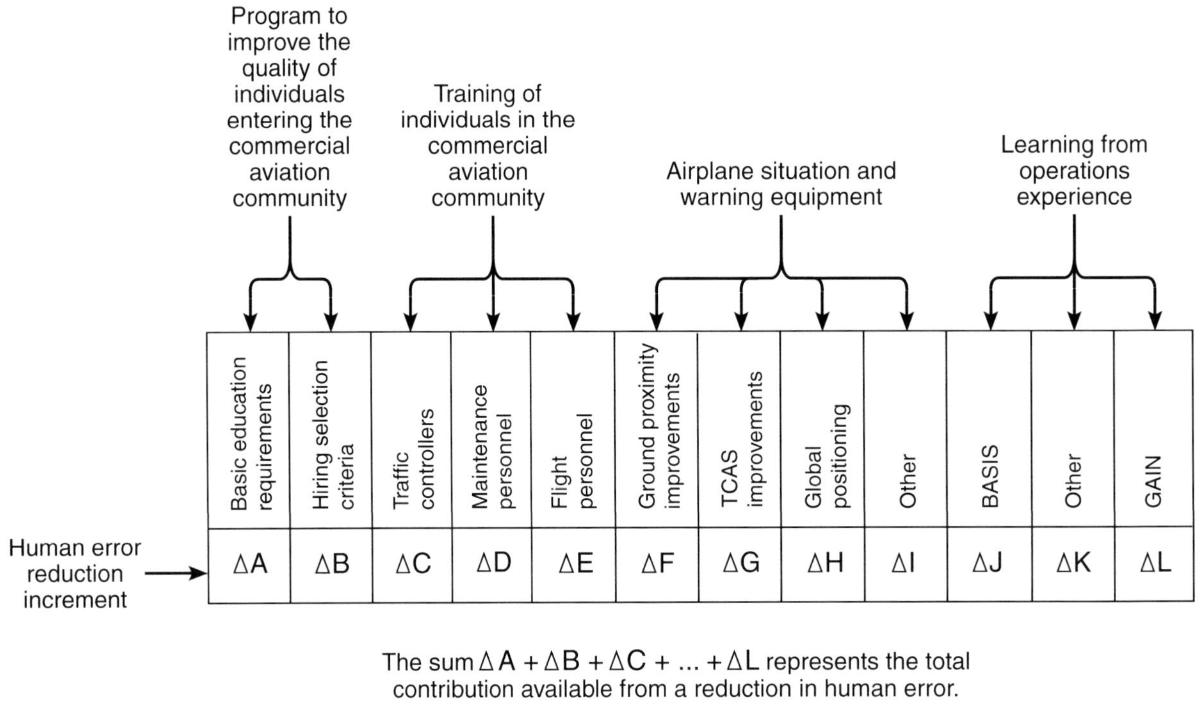

FIGURE 5-2 Current initiatives to reduce human error contributions to accidents/incidents.

- A video tape, "Every Day: A Programme about Error Management in Aircraft Maintenance," developed and published by the International Federation of Airworthiness (1997), features Professor James Reason and reviews human factor issues related to maintenance.

Additional work in fields such as cognitive science and fundamental neuroscience is progressing rapidly and is likely to offer valuable insights in the near future. The potential benefits of relying more on cognitive science are explained by James Reason (1990) and, in a more philosophical sense, by David Chalmers (1996). Turning to cognitive science to improve the understanding of issues associated with situational awareness has two major advantages. First, it should encourage the development of processes and systems that would improve the selection and presentation of necessary information, assigning to automated systems the tasks that systems do best and allowing people to continue doing the tasks that people do best. Second, it should help define the type of automation that can reduce the workload of flight crews and air traffic controllers in the crucial moments when a situation must be assessed quickly and accurately.

IMPROVING THE EFFECTIVENESS OF THE FAA'S HUMAN FACTORS PROJECTS

Harnessing the growing body of human factors knowledge will enhance the FAA's efforts to reduce the number of incidents and accidents by reducing human error and improving the ability of flight crews and other personnel to prevent accidents associated with other causes. That is one of the tasks of the FAA's Human Factors Study Group. This group appears to be reasonably well coordinated with the JAA (Joint Aviation Authorities) Human Factors Study Group and will operate indefinitely. Close coordination between these two groups is important in an environment that is becoming increasingly aware of the value of international harmonization of airworthiness standards and procedures. Coordinating the work of both groups with similar study groups sponsored by ICAO and other certifying authorities would also be worthwhile.

The membership of the FAA Human Factors Study Group should be reviewed and adjusted, if necessary, to ensure that it has strong representation from the fields of cognitive science and basic neuroscience. Strong representation in these areas would help the study group form a cohesive framework for understanding the very large number of human factors studies that are now under way, and it would enhance the ability of the group to recommend actions based on the results of these studies. Some of these studies are associated with enhanced ground proximity warning systems, improved traffic collision avoidance systems, and other aspects of developing crew-centered cockpit designs.

Finding 5-1. Maintaining situational awareness is the key to preventing the vast majority of serious incidents and accidents associated with human error.

Major Recommendation 4. The FAA should support and accelerate efforts (1) to define the minimum data required by the flight crew to maintain adequate situational awareness during all phases of flight and reasonable emergency scenarios and (2) to determine how this data can be presented most effectively.

Recommendation 5-1. The FAA should ensure that its human factors projects, especially the FAA Human Factors Study Group, include strong representation in the fields of cognitive science and basic neuroscience.

Recommendation 5-2. Advances in understanding human factors should be quickly applied to the key task of reducing the role of human errors in incidents and accidents, particularly with regard to improving the situational awareness of operational personnel and improving the effectiveness of maintenance personnel. The FAA should strongly support its Human Factors Study Group and other projects that contribute to this task.

REFERENCES

Chalmers, D. 1996. The Conscious Mind. Oxford, U.K.: Oxford University Press.

FAA (Federal Aviation Administration). 1997. Human Factors Guide for Aviation Maintenance. Washington D.C.: Office of Aviation Medicine, FAA.

FAA. 1996. The Interfaces Between Flight Crews and Modern Flight Deck Systems. Report of the FAA Human Factors Team. Washington, D.C.: FAA.

FAA. 1995. Proceedings of the FAA Workshop on Flight Crew Accident and Incident Human Factors. Washington, D.C.: Office of System Safety, FAA.

IFA (International Federation of Airworthiness). 1997. "Every Day: A Programme about Error Management in Aircraft Maintenance." Sussex, U.K.: IFA.

NASA (National Aeronautics and Space Administration). 1996. Principles and Guidelines for Duty and Rest Scheduling in Commercial Aviation. Washington, D.C.: NASA.

NTSB (National Transportation Safety Board). 1992. Flight-Crew-Involved Major Accidents of U.S. Air Carriers, 1978 through 1990. Washington, D.C.: NTSB.

Proceedings of the Ninth International Symposium on Aviation Psychology. April 28–May 1, 1997, Columbus, Ohio. Ohio State University: Aviation Psychology Laboratory.

Reason, J. 1990. Human Error. Cambridge, U.K.: Cambridge University Press.

6

Barriers

A number of barriers, both internal and external, will make it difficult to implement the recommended safety management process. The FAA has accepted the challenge of making aggressive improvements in aviation safety over the next 10 years. This goal cannot be achieved unless the FAA's efforts to improve the safety management process include dedicated efforts to overcome these barriers.

Major Recommendation 5. In order for AIR to contribute as much as possible to improvements in aviation safety, the FAA—in partnership with industry, Congress, the Department of Transportation, and other involved parties—must take aggressive action to overcome barriers associated with the following:

- external pressures and influences on the FAA
- coordination and communications within the FAA
- legal issues
- the rulemaking process
- the economic impact of proposed changes to the safety management process

EXTERNAL PRESSURES AND INFLUENCES

Highly publicized accidents are often caused by factors not associated with the greatest aviation hazards. Unfortunately, time and patience are required to determine the causes of any accident and to understand how to prevent a reoccurrence. Nevertheless, the FAA typically comes under considerable political pressures to do something immediately, even if it is based more on speculation than on facts. These pressures have sometimes caused the FAA to take corrective action that is otherwise unjustified.

Attention should be focused on the highest priority problems, as determined by accident and incident data—not by premature or politicized conclusions about the cause of the most recent major accident. The committee understands that major airline accidents naturally attract a great deal of attention from the media and others and agrees that these accidents deserve intense investigation to identify cause factors and appropriate corrective action. These investigations, however, should be structured in a way that does not interfere with other important safety projects. Implementing the recommended safety management process would help by reducing the number of accidents.

The time pressures typically associated with a "publicly investigated" accident may limit the effective use of risk analysis tools to identify cause factors and the most appropriate solutions. Public pressure may also result in action that is not supported by technical data and may be less effective in the long run than taking the time to develop a more effective response. For example, the secretary of transportation may dictate or industry may request an immediate AD to resolve the controversy associated with a particular accident and to deal with the bad publicity. In one case, political pressure following an in-flight explosion prompted the FAA soon afterwards to mandate aggressive new security measures because the accident was initially assumed to be a terrorist act, which was not the case. As a result, airline costs were increased unnecessarily and levels of service were reduced with no measurable improvement in safety. The crisis-like atmosphere and the public search to find someone to blame for an accident also tends to create an atmosphere of mistrust and animosity that contributes to public questions about aviation safety and whether industry and government are acting in good faith, even though aviation is in fact the safest mode of transportation in the United States.

The NTSB has the sole responsibility for determining the probable cause of an accident. The NTSB may make safety recommendations during the course of an investigation that it later concludes are unrelated to the probable cause. The FAA has a statutory responsibility to participate in accident investigations and to order corrective action it determines is necessary to prevent a reoccurrence. During the long interval between the occurrence of an accident and the time the NTSB makes a formal finding of probable cause, conflicting views often develop among the NTSB, FAA, and other parties on what corrective action should be taken. These conflicts can be healthy when they test opinions and ensure that the most effective preventative measures are implemented. However, sometimes one or more parties try to shape public opinion in its favor. This can create additional pressures that are counterproductive and contrary to the public interest.

In response to NTSB recommendations that require regulatory action, the FAA is required by statute and executive order to complete cost-benefit analyses. These analyses compare the financial costs of implementing the recommended corrective action with the monetary value of expected benefits. The Department of Transportation specifies the monetary value ($2.7 million) of each life that would be saved by the recommended action. Until recently, if the FAA could not document that benefits exceeded costs, it was prohibited from adopting the recommendation. Even now, a recommendation is rarely implemented if its benefits cannot be shown to exceed its costs. The NTSB has no similar constraints on making safety recommendations.

For major regulatory actions the Office of the Secretary of Transportation and/or OMB verify the FAA's cost-benefit analyses. However, if the FAA determines that recommended actions will not be cost effective and decides, or is directed, not to implement an NTSB recommendation, the FAA may be harshly and publicly criticized for failing to take action. This criticism rarely seems to recognize the different roles of the NTSB and the FAA or the statutory requirements that the FAA must satisfy to take regulatory action (see Chapter 2).

The committee recognizes and endorses the NTSB's status as an independent accident investigatory agency. However, greater harmony between the FAA and NTSB would likely improve the ability of both agencies to address the most significant risks to aviation safety and increase the confidence of Congress and the public in the safety of the air transportation system.

Finding 6-1. Following some highly publicized accidents, there is a technically unjustified loss of public confidence, which leads to political pressure and a counterproductive atmosphere of crisis management in the FAA that interferes with ongoing efforts by government and industry to improve aviation safety.

One approach for addressing this issue would be for the Department of Transportation, FAA, and NTSB to develop a joint process that reduces interagency disagreements about important safety recommendations, while protecting the NTSB's role as an independent agency and the FAA's ability to comply with the requirements of the rulemaking process. This process could include the following elements:

- establishment of a senior interagency communications or safety management board that would coordinate the government's public comments on aviation disasters
- improved communications between the FAA and NTSB, perhaps through combined analysis meetings and short-term personnel exchanges
- establishment of a combined technical team to assess how NTSB recommendations should be disposed of in the context of FAA rulemaking procedures and requirements

Recommendation 6-1. As a first step towards reducing the negative impact of external pressure on the safety management process, the FAA should work with other responsible agencies to educate the public more fully about ongoing efforts to improve aviation safety. Fully addressing this issue is likely to require major organizational changes, such as the establishment of a senior interagency communications or safety management board, that were beyond the scope of this study.

COORDINATION AND COMMUNICATIONS

Many of the individual organizational elements within the FAA enjoy considerable autonomy over their assigned areas of responsibility but lack an effective means of communicating with each other and resolving differences. As demonstrated by the long time it typically takes to make regulatory changes, the lack of communication sometimes prevents the FAA from taking timely action to improve overall safety or operational efficiency. More effective communications—within AIR and between AIR and other FAA offices, such as the Flight Standards Service—would considerably improve the aircraft certification safety management process. Consider the following two examples.

This report recommends that the FAA eliminate internal efforts to collect and store data for aircraft manufactured by companies that agree to assume this function. Currently, data is collected by many different offices within AIR and other parts of the FAA. To be most effective, defining how the FAA can increase its reliance on external databases should involve each office that currently collects and maintains data. This would ensure that the new approach will meet FAA data requirements and avoid inappropriate duplication or gaps in data collection and storage. Such an approach, however, would require effective communications and strong leadership so that efforts to obtain the support of involved parties would not prevent timely action.

The relationship between AIR and the Flight Standards Service is also important. AIR has primary responsibility for developing and administering aircraft airworthiness standards and issuing related ADs. The Flight Standards Service has primary responsibility for approving operational procedures and minimum maintenance requirements and for checking that operators comply with airworthiness standards. Each organization has knowledge, expertise, sources of information, and channels of communication with outside organizations that are more fully developed in some areas than others. Close coordination between AIR and the Flight Standards Service helps ensure that solutions are based on the best information available and that corrective action is disseminated as effectively as possible.

Recommendation 6-2. The FAA should develop a process to facilitate communications and improve coordination among offices within AIR and between AIR and the Flight Standards Service. For example, the Associate Administration for Regulation and Certification could establish a central coordinating office to facilitate the exchange of continued airworthiness information within the FAA and the dissemination of complete and consistent information to industry.

LEGAL ISSUES

Although the committee was not constituted to address legal issues in detail, it identified three legal issues that could delay full and effective implementation of the recommended safety management process. These issues are associated with the potential for public disclosure of sensitive information under the Freedom of Information Act, the possibility of regulatory enforcement against individuals or companies who voluntarily disclose unfavorable safety data, and increased exposure to legal liability arising from the litigation discovery process in an environment where more data is collected, stored, and shared. The recommended safety management process advocates the voluntary participation of airlines, manufacturers, and other stakeholders during the long period it would take to develop new regulations mandating compliance. Thus, resolving industry's widely held concerns about legal issues is essential. Even a mandatory system would rely heavily on the good faith of all participants, and good faith is not likely to last very long if pilots, operators, and manufacturers are punished for voluntarily drawing attention to safety problems that would otherwise go unnoticed.

Public Disclosure

Under the recommended safety management process, the FAA would have access to a large amount of safety data generated and collected by industry. This access would enable the FAA to oversee the recommended safety management process. However, the safety data, which are likely to be considered sensitive and proprietary by manufacturers and operators, could be subject to disclosure outside the government under the Freedom of Information Act. Safety data can be disclosed to the FAA under restrictive legends and/or confidentiality agreements, but there is no guarantee that the data would be uniformly protected because disclosure determinations under the Freedom of Information Act are subject to judicial discretion and conflicting rulings by various courts. Moreover, once data have been shared with competitors, they can no longer be considered confidential commercial or financial information. The Federal Aviation Reauthorization Act of 1996 addressed this concern by changing Title 49 of the United States Code. This change directs the FAA not to disclose voluntarily-provided safety- or security-related information "notwithstanding any other provision of law" if the FAA administrator finds that

> (1) the disclosure of the information would inhibit the voluntary provision of that type of information and that the receipt of that type of information aids in fulfilling the Administrator's safety and security responsibilities; and (2) withholding such information from disclosure would be consistent with the administrator's safety and security responsibilities.

The FAA is in the process of changing the FARs to implement this legislation.

Enforcement Action

The FAA's current policy is to not use FOQA (flight operations quality assurance) data provided by industry voluntarily as a basis for enforcement action, and that policy is being codified by a change in the FARs. This change is not likely to apply to enforcement action associated with other types of information that may be voluntarily provided to the FAA.

Discovery

Legal discovery is a more difficult matter. Industry may be reluctant to increase the amount of safety data it generates and disseminates voluntarily because of concerns that the information will be subject to discovery in civil lawsuits. Moreover, increasing the volume of information could lead to increased potential liability, particularly in areas where liability may not currently exist (i.e., where dissemination of safety information notifies a manufacturer or operator of potential safety problems that were unknown to it). Industry's concerns in this regard may be addressed in part if identifying information is redacted from performance and safety data prior to dissemination. The level of detail retained in the shared data would need to be weighed against the proprietary and privacy rights of the organizations providing the data and the need for a database detailed enough to support a robust risk analysis and risk management process. Even then, the unredacted performance data would still potentially be subject to civil discovery.

The possibility of discovery is a disincentive for manufacturers and operators to participate in the safety management process. The committee believes that the public interest would be served and safety would be enhanced by legislative action to address this disincentive to the voluntary sharing of safety data among operators and manufacturers.

Section 701(e) of the Federal Aviation Act of 1958 and section 304(c) of the Independent Safety Board Act of 1974 specify that NTSB accident reports are not admissible in civil actions. The Congress should consider legislation that would also limit the use of manufacturers' and operators' safety analyses in civil suits against the company that supplied the information.

Recommendation 6-3. The FAA should initiate regulatory action, legislative action (through the Congress), and/or letters of understanding with manufacturers, operators, pilot organizations, and others to serve the public interest and improve safety by encouraging the voluntary sharing of safety data, including data generated by industry safety analyses. This may involve limiting enforcement action based on voluntarily shared data and protecting such data from release to other parties.

RULEMAKING PROCESS

The FAA's rulemaking process, which is described in Chapter 2, quite often takes 5 to 10 years to revise airworthiness standards. Rulemaking projects are sometimes completed much more quickly, especially in the case of highly publicized or emergency safety actions. However, many worthwhile, safety-related rulemaking projects languish for years. The barriers to quick action are many and have a considerable history. For example, the current rulemaking processes is complex and not enough staff hours are dedicated to rulemaking in AIR and other offices, such as the Office of Rulemaking, Office of Policy and Plans, and Office of the Chief Counsel. Significant delays also have occurred during reviews of NPRMs (notices of proposed rulemaking) and final rules by the Office of the Secretary of Transportation. Tremendous delays are caused by differing, strongly held points of view within the FAA, other federal agencies, and/or industry. In such cases, deferring action or referring a matter back to another office for reconsideration may be the easiest—but least productive—course of action.

The FAA is certainly not the only federal agency that has difficulty making timely changes to federal regulations, but delays in the federal aviation rulemaking process are a significant safety issue. In fact, some safety improvements are structured to circumvent the rulemaking process because of the delays it would otherwise take to implement them in regulatory form. The FAA may issue advisory circulars, which provide nonmandatory guidance, or rely on draft documents to provide interim guidance during the years it takes to issue final policies or regulations. In other cases, industry may delay safety improvements until the FAA decides if the proposed action is a sufficient response to a particular concern. Industry is often reluctant to make safety improvements, such as expensive aircraft modifications, without assurance from the FAA that new rules will not require undoing or redoing the modifications. Individual operators who decide to implement safety enhancements voluntarily may place themselves at a competitive disadvantage. The public is not well served by a process that pressures operators to delay implementation of safety enhancements.

The Federal Aviation Reauthorization Act of 1996 addressed rulemaking delays by requiring the FAA to issue a final rule within 16 months after the close of public comments on an NPRM. However, it quite often takes several years for an NPRM to be generated and approved by the ARAC (Aviation Rulemaking Advisory Committee); to be revised and approved by cognizant officials in the FAA, Department of Transportation, and OMB; and to be issued for public comment. The 1996 act does not limit how long this part of the rulemaking process may take. Neither does the act allocate additional resources to rulemaking or waive any regulatory or legislative constraints on the rulemaking process. The FAA has not yet determined how to meet the timeliness requirements of the 1996 act for the large number of rulemaking projects that are typically under way at any one time, nor has it determined how to significantly accelerate the preparation and issuance of NPRMs. Reassigning the duties of current FAA personnel could reduce time delays, and establishing ADOs (approved design organizations) in accordance with Major Recommendation 3 could help by reducing the number of personnel required to support certification activities.

The FAA established the ARAC and its industry working groups to facilitate the rulemaking process. Each of the numerous working groups is assisted by FAA technical experts, cost-benefit analysts, and legal representatives. The ARAC was expected to develop all of the documents necessary to implement its rulemaking recommendations, including NPRMs, cost-benefit analyses, and advisory circulars describing how to implement new or modified regulations. When the ARAC was chartered in 1991, expected benefits included the following:

- taking advantage of industry's domestic and international technical expertise and experience to prepare better rules
- resolving controversies in an open forum prior to the formal rulemaking process, thereby shortening the time required for the disposition of comments elicited by NPRMs
- broadening participation in rulemaking by the public, industry, and the aviation authorities of other countries (thereby facilitating the international harmonization of aviation safety regulations)

The ability of the ARAC to shorten the rulemaking process has been limited for the following reasons:

- The ARAC itself has become large and unwieldy, with representatives of about 70 organizations. The ARAC also has about 50 working groups (each working on multiple subjects) with a total of about 400 participants. Reaching consensus is time consuming, often taking years, because of the large number of participants and because of the turnover of participants resulting from the protracted duration of individual projects.

- ARAC recommendations are subject to the same lengthy rulemaking process as proposed rules developed without the assistance of the ARAC. The net result: even after the ARAC has invested years reaching consensus, it may take several more years for the FAA to act on a proposal.
- Proposals for many worthwhile and noncontroversial rule changes are backed up within the FAA behind higher priority rule changes that have been delayed because of political sensitivities. As of January 1998, more than 300 rules and rule proposals were awaiting action by the FAA.
- Although the FAA participates in the ARAC process as an observer, industry perceives that many FAA participants either drive committee deliberations toward predetermined positions, which contributes to delays in reaching consensus, or do not accurately portray FAA technical and policy concerns with proposed actions.
- Rulemaking documentation produced by the ARAC often requires extensive reworking after it is submitted to the FAA, which increases delays. Some industry representatives believe this results from having FAA participants with limited expertise and authority who cannot effectively represent FAA positions. Conversely, some FAA personnel believe the problem is the lack of expertise of industry participants. To some extent, both of these views are correct. Doubts about the ARAC's timeliness and effectiveness reduce the incentive for both the FAA and industry to assign more skilled and experienced personnel to ARAC activities.

Recommendation 6-4. Efforts to reform the ARAC should (1) establish more timely and effective processes and (2) encourage the assignment of industry and FAA personnel who have the expertise to develop well written NPRMs and final rules and the influence necessary for building broad support for documents approved by the ARAC.

In August 1996, the FAA's Office of Rulemaking initiated the Rulemaking Process Reengineering Project, which included participation by other FAA offices concerned with the efficiency of the FAA rulemaking process. The FAA administrator reviewed the results of this project and approved implementation of its recommendations starting in early 1998. These recommendations are intended to accomplish the following:

- Improve FAA management of the rulemaking process, including management of ARAC activities.
- Improve the quality of the rulemaking products at all stages.
- Facilitate early resolution of differences among major stakeholders.

This project was limited to a review of internal rulemaking procedures. The committee did not have an opportunity to assess the project's recommendations or the extent to which they are likely to expedite the rulemaking process.

Finding 6-2. It quite often takes 5 to 10 years to issue new regulations or modify existing regulations. This is an important safety issue because it constrains the ability of the rulemaking process to improve aviation safety. The FAA is in the process of reforming its internal procedures, including ARAC procedures. This is a positive first step, but much more needs to be done in this area, and time is of the essence.

Recommendation 6-5. The FAA should make the rulemaking process substantially more responsive by convincing the Department of Transportation, other executive branch agencies, and Congress to modify legislation, directives, and regulations to allow major changes in the current process.

ECONOMIC IMPACT

Over the long term, more comprehensive data collection and analysis systems are expected to reduce costs and improve safety. The air transport industry is highly competitive, however, and this natural competitiveness could present a barrier to the voluntary sharing of data required to implement the recommended safety management process. For example, the cost of establishing the recommended database management and risk analysis systems could be perceived as a barrier. The costs and associated benefits must therefore be thoroughly evaluated by the FAA in cooperation with industry, and the FAA should provide financial support for pilot programs to validate costs and demonstrate the effectiveness and practicality of implementing procedures. This approach would be similar to the one that the FAA is using to foster the spread of FOQA systems among U.S. operators, in which the FAA funds pilot projects by several U.S. airlines.

A similar problem will occur as the safety management process becomes more proactive, identifying safety actions that should be taken to eliminate risks before they cause an accident. Manufacturers and operators bear the cost of making safety improvements, and they will not support improvements unless the identified risks are credible and the corrective action seems reasonable in terms of effectiveness, direct costs, and indirect costs, including disruptions in operations and damage to their corporate reputations. It is imperative that any new risk management system have enough credibility to justify making increased financial demands on industry, especially if the financial demands could reduce competitiveness. This process may be simplified if enhanced data collection, database management, and analysis systems reduce operating costs. For example, FAA-funded pilot projects have shown that FOQA systems can reduce airline operating costs more than enough to pay for themselves.

Recommendation 6-6. The FAA should work with industry to develop confidence in the cost-benefit analyses used to justify changes in the safety management process. The FAA should also subsidize pilot projects by operators and manufacturers to validate the cost effectiveness of new systems for data collection, database management, and analysis.

7

Small Airplanes and Rotorcraft

INTRODUCTION

U.S. civil aviation includes about 180,000 general aviation aircraft, the vast majority of which are small airplanes (with a maximum takeoff weight of less than 12,500 pounds). About 5,100 are rotorcraft. By comparison, scheduled air carriers (FAR Parts 121 and 135) operate more than 7,000 aircraft, including about 5,000 large turbojet aircraft (the rest are turboprop airplanes, piston-engine airplanes, and rotorcraft). The total fleet of general aviation aircraft accumulates about 24 million flight hours annually, compared to 14 million flight hours for large air carriers and 5 million flight hours for commuter airlines and air taxis. Similarly, the number of small airplane and rotorcraft operators in the United States (approximately 100,000) is also much larger than the number of large airplane operators (just 153 for FAR Part 121 air carriers), and the former encompass a much wider range of pilot experience and skills than the latter (GAMA, 1997; FAA, 1996a, 1996b).

Some corporate and commercial operators of small airplanes and rotorcraft use safety procedures and techniques very similar to those of large airlines. For example, the aviation departments of many Fortune 500 corporations have airplane safety records comparable to those of large airlines. Rotorcraft operators supporting offshore oil production and corporations that provide aircraft for executive transportation also tend to have relatively sophisticated operations. However, most operators of small airplanes and rotorcraft have just a few aircraft and do not have sophisticated institutional safety programs. For example, 75 percent of the operators who belong to the Helicopter Association International operate fewer than five helicopters, and 39 percent operate only one. On the other hand, the 13 largest U.S. carriers, with turbojet fleets ranging from 150 to almost 700 aircraft, have an average of more than 300 turbojet aircraft. The 62 other carriers that operate turbojets, with fleet sizes ranging from 1 to 35 aircraft, have an average of 10 turbojet aircraft (FAA, 1994).

Small airplanes and rotorcraft operate in a much broader spectrum of functional modes than most large airplanes. For example, small airplanes and rotorcraft operate as air taxis, corporate aircraft, business aircraft, personal aircraft, and instructional aircraft. Other roles include sightseeing, pipeline patrol, law enforcement, emergency rescue, scientific experimentation, transport of external loads, crop dusting, and firefighting. Operational cycles are also very different. A typical small airplane or rotorcraft is in the air many fewer hours per year than a typical large transport—an average of 140 hours for small airplanes and rotorcraft, compared to 3,000 to 3,500 hours for large transport airplanes operated by major air carriers. Small airplanes and rotorcraft also operate out of many more airports and landing areas than large airplanes, and many of these lack control towers and other landing and takeoff aids. Small aircraft, rotorcraft, and large transport airplanes do share much of the same airspace and use many of the same facilities, however. Thus, despite their differences, it is essential that systems and procedures allow them to operate together safely (GAMA, 1997; FAA, 1996a, 1996b).

The safety management process for small airplanes and rotorcraft must be flexible enough to accommodate the diverse nature of these communities, and this is likely to be a difficult challenge. Final accident investigation reports for small airplanes and rotorcraft (as with large transports) show that the majority of accidents are attributable to human error, and the small role played by aircraft system malfunctions indicates that the current aircraft certification and continued airworthiness process is working well.

This chapter deals with the differences in the safety management processes applicable to a typical transport airplane, such as a large passenger jet operated by a major airline, and a typical small airplane, such as a small general aviation aircraft or helicopter owned by an individual or business that may not own any other aircraft. The committee acknowledges, but does not specifically address, additional considerations raised by less common—but hardly unusual—

situations where small commercial air carriers operate small aircraft or a large corporation operates a fleet of business jets larger than the aircraft operated by many commuter airlines.

SAFETY AND SAFETY MANAGEMENT PROCESSES

Because of the similarities in circumstances and conditions relating to accidents involving large airplanes, small airplanes, and rotorcraft, the committee believes the recommendations that appear elsewhere in this report, which are focused on accidents and incidents involving large airplanes, are generally applicable to small airplanes and rotorcraft. The differences that do exist, however, indicate that different means should be used to carry out many of the committee's recommendations.

Causes of Accidents and Incidents

Uncertainties regarding two important factors degrade the accuracy of accident statistics for small airplanes and rotorcraft: (1) the appropriate categorization of accidents according to type of operation and (2) the accuracy of flight hour information for each type of operation.

The NTSB is unable to conduct detailed investigations into most accidents involving small airplanes and rotorcraft because of the large number of such accidents and the limited resources that the NTSB has available for this task. The NTSB conducts many of these investigations using telephone inquiries with on-scene personnel instead of dispatching field investigators to the site. If the pilot is killed and there are no other survivors (or no one else was on board), it is sometimes very difficult to accurately determine what events preceded the accident. Because of the great variety of small aircraft and small aircraft operations, little or no mechanical or system trend analyses are performed to better understand the underlying causes of small aircraft accidents.

Unlike most operators of transport airplanes, most operators of small aircraft are not required to report operational statistics, such as flight hours, to the FAA or any other government agency. Flying hours are presently estimated by the FAA using statistical forecasting techniques from its "General Aviation and Air Taxi Activity and Avionics Survey," which is distributed annually to a sample population of aircraft owners. Responses are not mandatory, and small operators engaged in varied operations may not have accurate records of flight hours broken down by type of operation. In addition, NTSB statistical summaries on general aviation operations and accident rates, which are derived from the same FAA survey, take nearly a year to finalize and are only available on an annual basis. The resulting lack of precision in estimated flight hours results in accident rate statistics that are equally imprecise. More precise calculations and timely dissemination of accident statistics would be helpful to understand current trends and the effectiveness of accident prevention measures.

Although there were fewer accidents in 1996 involving small airplanes and rotorcraft as a group than in the several prior years, small airplanes and rotorcraft (like large airplanes) seem to be experiencing a relatively stable fatal accident rate. With the number of flying hours projected to increase over the next decade, small airplanes and rotorcraft are expected to experience an increase in the total number of fatal accidents.

As with scheduled air carriers, final accident investigation reports for small airplanes and rotorcraft show that the majority of accidents are attributable to human error. The incomplete understanding of many human error-related accidents emphasizes the need for continued work in this area, as recommended in Chapter 5. Another similarity shared by large and small aircraft is the small role that aircraft system malfunctions play in accidents, which indicates the ongoing effectiveness of the current type certification and continued airworthiness process.

Capabilities of Manufacturers and Operators

The safety management process for the small airplane and rotorcraft communities must be flexible enough to accommodate the diverse nature of these communities, and this is likely to be a difficult challenge. Safety management processes for small airplanes and rotorcraft must overcome challenges associated with a much greater assortment of aircraft designs, more varied operational roles, and a much larger number of operators than those of large airplanes. In most cases, these differences are inherent and unavoidable. For example, large transport airplanes carry sophisticated flight management systems and safety devices, which have helped them achieve a much lower accident rate than small airplanes and rotorcraft. However, the cost of these systems exceeds the total value of many small airplanes, and the systems would be impractical to install on small airplanes or rotorcraft because of configuration limitations (weight, volume, electrical power, etc.).

Very few operators of small airplanes and rotorcraft have the resources to establish flight operations, aircraft maintenance, or data analysis comparable to those of major airlines. Many rely almost exclusively on other organizations, such as the FAA, manufacturers, repair stations, individual licensed mechanics, and/or professional organizations, to provide these resources. In particular, many small operators rely on the FAA to tell them (in the form of an AD) when special action is needed to correct unexpected safety deficiencies in their aircraft. Yet it is often difficult for the FAA to obtain comprehensive safety-related feedback upon which to base ADs because the applicable regulations (FAR Parts 61, 63, 65, 91, 133, and 137) do not require most operators of small airplanes and rotorcraft to report safety hazards.

Currently, many manufacturers of small airplanes and rotorcraft cooperate with the FAA and other aviation organizations to provide a variety of training and accident prevention programs. However, many models of aircraft are no longer supported by manufacturers because the manufacturers have gone out of business or the aircraft are so old that manufacturers no longer produce parts for them.

Data Collection, Database Management, and Risk Analysis

AIR can fulfill its mandate to maintain the airworthiness of aircraft only if it has access to valid information about service difficulties as they develop. The data collection, database management, and risk analysis methods recommended in Chapter 4 rely heavily on manufacturers and operators to provide these services. This approach cannot be applied directly to the small airplane and rotorcraft industries because of previously noted limitations on the capabilities of most operators and manufacturers of small airplanes and rotorcraft, particularly in cases where manufacturers are no longer in the aircraft manufacturing business. In addition, AIR has much less regulatory control over most operators of small airplanes and rotorcraft than it does with a typical transport airplane operator.

Nevertheless, the small airplane and rotorcraft industries have long recognized the need to know "what's going wrong" before adverse situations develop into an accident. Major associations of operators and manufacturers of small airplanes and rotorcraft have tried to address this problem. For example, the Helicopter Association International (the membership of which includes manufacturers and operators) has developed, with the cooperation and support of AIR's Rotorcraft Directorate, a computerized system for collecting data on operational problems from helicopter operators. The information from this system is then made available to the respective manufacturers.

The helicopter industry (particularly the European industry) also has been developing health and usage monitoring systems to monitor the unique characteristics of helicopters. Although the evolution of a practical system has been hindered by variations in technical expertise among users, designers, and regulatory authorities, AIR's Rotorcraft Directorate is attempting to resolve these difficulties in conjunction with JAA and industry. Meanwhile, some of the larger helicopter operators have adopted health and usage monitoring systems on their own initiative and appear to be encouraged by the results. AIR is also trying to improve its ability to manage and analyze available accident and incident data through its Aviation Safety Management Program and other programs.

Major Recommendation 6. Plans to implement the recommended safety management process within the small airplane and rotorcraft communities should be developed in cooperation with small airplane and rotorcraft operators, manufacturers, and associations of operators and manufacturers. The FAA should establish cooperative agreements that define the roles of individual operators, individual manufacturers, their associations, and AIR. These agreements should define the following:

- responsibilities of operators for submitting data
- responsibilities of operators, manufacturers, associations of operators and manufacturers, and AIR for data collection, database management, risk analysis, risk management/action, and monitoring effectiveness
- processes for the routine exchange of data and risk analysis results between operators, manufacturers, associations, and AIR to facilitate effective risk management/action
- a publicity program to inform the small airplane and rotorcraft communities of the new safety management process

ADDITIONAL SMALL AIRPLANE CONCERNS

As already noted, small airplanes include a broad spectrum of airplane designs, operators, and missions. Although small airplane accidents cause more fatalities than large aircraft accidents, individual accidents are rarely newsworthy, and accident prevention for this segment of the air transportation system receives secondary attention from the media, NTSB, and FAA. In this context, it is difficult to develop a comprehensive understanding of how and why certain kinds of accidents occur and how to prevent them.

Although human factors are clearly the leading cause of small airplane accidents, NTSB accident reports often provide only sketchy details about the human factors leading to an accident. In addition, the NTSB only performs field investigations of approximately 20 percent of general aviation accidents. As a result, the examination of aircraft systems and other physical evidence is sometimes incomplete, making it difficult to identify trends and implement broad corrective action in a timely fashion. Increasing the number of NTSB field investigations of small airplane accidents and the amount of human factors information gathered during these investigations would help address this problem.

General aviation involves many dissimilar segments, with widely differing aircraft designs, regulations, and pilot capabilities. For example, general aviation aircraft—of which small airplanes are by far the largest part—are involved in nearly 2,000 accidents per year, with an estimated accident rate of 8.06 per 100,000 flight hours and a fatal accident rate of 1.51 per 100,000 flight hours. This is 25 to 50 times higher than the corresponding values for the corporate aviation segment of the general aviation community (0.14 accidents per 100,000 flight hours and 0.06 fatal accidents per 100,000 flight hours) (NTSB, 1996). In fact, accident rates for corporate aircraft are comparable to accident rates for Part 121 air carriers (0.21 accidents per 100,000 flight hours and 0.035 fatal accidents per 100,000 flight hours) (FAA, 1994). Corporate aircraft are flown by professional pilots, and the disparity in accident rates indicates that many small airplane accidents may result from pilots who are at risk because they lack the piloting skills or experience to identify a problem, properly evaluate the risk it poses, and take appropriate action before it is too late. An important safety challenge is to identify these pilots and improve their decision-making skills for situations in which they are likely to be at greatest risk.

Recommendation 7-1. The FAA should conduct separate safety assessments for each segment of the general aviation community to identify the continued airworthiness problems of greatest significance as a function of the type of operation, class of aircraft, and experience level of the pilots.

ADDITIONAL ROTORCRAFT CONCERNS

Rotorcraft—for which most experience to date comes from helicopters—have many of the characteristics and use many of the technical developments as fixed-wing aircraft. However, there are also important differences between rotorcraft and fixed-wing aircraft in general, and between rotorcraft and large transport airplanes in particular. Because of these differences, special efforts—including some procedural changes—will be needed to implement the recommended safety management process in a way that accommodates unique rotorcraft safety considerations and the capability of the rotorcraft industry to identify safety issues based on operator reports.

Certification of Surplus Military Helicopters

The certification of military surplus helicopters is currently an area of particular concern to the rotorcraft industry. In 1995, the U.S. Army decided to sell or give away about 3,000 surplus helicopters over the next several years (and, presumably, associated surplus spare and replacement parts). This decision caused a great deal of concern within the rotorcraft community for several reasons:

- the potential economic impact that the sale of inexpensive surplus helicopters and parts could have on the market for newly manufactured equipment (an issue that is not relevant to this study)
- the potential safety impact of using surplus military parts on helicopters built to civil designs
- uncertainties about how accurately the certification process would assess the safety of using helicopters in civil operations that were not necessarily designed, manufactured, operated, or maintained in accordance with civil airworthiness standards

In many cases, military standards and practices would meet or exceed civil airworthiness standards. But this is not true in all cases, and defining a certification process that would protect public safety—without unduly impairing the ability of prospective purchasers to make appropriate use of their aircraft—is a complicated problem.

FAA certification regulations define requirements for converting surplus military aircraft to civil use. Procedures for fixed-wing aircraft were first developed immediately after the end of World War II, when there was a flood of surplus military aircraft, many ex-military pilots were searching for ways to stay in the aviation business, and most manufacturers were configured to produce military aircraft. At that time, the Civil Aeronautics Administration (the predecessor to the FAA) established a test program to quickly assess the characteristics of available military aircraft. Some of these aircraft—particularly aircraft that had been purchased off-the-shelf by the military—were, in fact, built to civil designs, and they were promptly absorbed into the airline fleets. Other aircraft were required to undergo a more lengthy certification process. These procedures have been modified many times over the years and now provide for civil certification of ex-military helicopters.

Current FAA regulations identify two methods by which aircraft manufactured in accordance with the requirements of, and accepted for use by, the armed forces of the United States can be certificated for civil use:

- Certification in the restricted category is possible for military surplus aircraft that have been modified for a limited number of specifically identified special purposes, will be operated only for those special purposes, and will not carry persons or property for compensation or hire (except for helicopters, which may be certificated to transport for hire objects carried as an external load).
- Certification in the normal or transport category is possible for aircraft that were designed and constructed in the United States and can be shown to comply with the standard FARs in force at the time the aircraft were accepted by the armed forces.

The current civil fleet of military surplus helicopters being operated today in the United States is composed primarily—if not entirely—of helicopters in the restricted category. Safety concerns for helicopters certificated in the restricted category relate to the following factors:

1. the physical condition of the aircraft
2. the degree to which military safety directives have been implemented
3. the completeness and accuracy of the aircraft's operation and maintenance records
4. military limits of operation and whether they will apply to civil operation

More than fixed-wing aircraft, the service life of many helicopter systems and components is strongly influenced by a history of severe use, such as frequent consecutive carriage of heavy loads, and it can be difficult to accurately assess the severity of use of an individual military helicopter. Thus, complete and accurate records of operation and maintenance are especially important for every ex-military helicopter being considered for civil certification.

Many military helicopters are variants of designs certificated for civil use and, because they generally look the same as their civil counterparts, the model designations can be easily confused. Even though the military and civil variants were designed and built by the same manufacturer, they may have been constructed under different quality control standards and to different design criteria and operational limits.

In 1976, the FAA, with the agreement of the Department of Defense, initiated a program to determine the feasibility of certificating surplus military aircraft through FAA Order 8130.6, which has since lapsed and been replaced by Order 8130.2C. As requested by the military, FAA inspectors examine each aircraft being released and validate its potential for civil certification. Chief among the conditions surveyed are items (1), (2), and (3), above, and the presence of the name plate from the original manufacturer. Many aircraft that undergo this examination are found to be unworthy of civil certification, although they may be suitable as a source of spare parts. Procedures and criteria for the release of spare parts by the military and guidelines for handling those parts are being negotiated by the FAA and the Department of Defense, and the FAA intends to issue an advisory circular on this topic.

Unfortunately, FAA certification offices have little specific guidance for evaluating applications for restricted category type certificates involving military surplus helicopters. Different aircraft of the same military model have received different restricted category type certificates from different FAA offices. In addition, applications for restricted certificates may receive less attention than standard certification programs because the aircraft are probably going to be used only for industrial work, which reduces the risk to the general public. The safety of the crew and third parties remains a valid concern, however.

Recommendation 7-2. AIR, in conjunction with the original equipment manufacturers of military surplus helicopters, should take timely action to define specific guidance for ACOs (aircraft certification offices) and industry to use in evaluating the airworthiness of military surplus helicopters in accordance with current regulatory standards.

REFERENCES

FAA (Federal Aviation Administration). 1994. 1994 Statistical Handbook of Aviation. Washington, D.C.: FAA.

FAA. 1996a. Aviation System Indicators 1996 Annual Report. Washington, D.C.: FAA.

FAA. 1996b. Aviation Safety Statistical Handbook. Washington, D.C.: FAA.

GAMA (General Aviation Manufacturers Association). 1997. General Aviation 1997 Statistical Handbook. Washington, D.C.: GAMA.

NTSB (National Transportation Safety Board). 1996. Annual Review of Aircraft Accident Data. U.S. General Aviation Calendar Year 1994. Washington, D.C.: NTSB.

APPENDICES

APPENDIX A

List of Findings and Recommendations

A complete list of the committee's findings and recommendations appears below in the order they appear in the body of the report.

CHAPTER 3

Causes of Incidents and Accidents

Finding 3-1. Safety management processes that focus on the primary causes of accidents are reactive and are unlikely to address some important cause factors adequately. Data from investigations of accidents *and incidents* are essential for planning proactive corrective action, which should address all important cause factors.

CHAPTER 4

Recommended Safety Management Process

The Major Finding. The recommended safety management process should improve the ability of the FAA/AIR, manufacturers, and operators to take corrective action based on assessments of incident data—before an accident takes place—and to set priorities based on current and future risk. However, the current process is already highly effective—as indicated by the small contribution of aircraft system malfunctions to the overall accident rate—and changes to the current system must be carefully structured to avoid unintended consequences that might reduce safety in some situations.

Major Recommendation 1. It is critically important that the FAA and AIR conduct business in a new fashion with regard to aircraft certification and continued airworthiness. As an essential first step, AIR should revise its budget and manpower allocations to better reflect its mission priorities, which are as follows:

1. continued airworthiness and other activities related to continued operational safety
2. rulemaking and policy development
3. certification

Major Recommendation 2. It is essential that the FAA improve its safety management process. The FAA should work with the operators and manufacturers of large transport airplanes and engines to define and implement a proactive process that includes the following elements and tasks:

Key Elements
- data collection
- database management
- risk analysis
- risk management/action
- monitoring effectiveness

Specific Tasks
- Manufacturers, with the advice and consent of operators and the FAA, should define data requirements and processes for sharing data. Comprehensive flight operations quality assurance systems similar to the British Airways Safety Information System (BASIS) should be used as a starting point.
- Operators should provide required data, as agreed upon.
- Manufacturers should solicit data from additional sources, such as the National Transportation Safety Board, International Civil Aviation Organization, and National Aeronautics and Space Administration, to augment the operational database.
- Manufacturers, with oversight from the FAA and the assistance of operators, as required, should collect, organize, and analyze data to identify potential safety problems.
- Manufacturers should recommend corrective action for potential safety problems and seek consensus by operators. The FAA should make sure that actions proposed

by manufacturers and operators will be effective, making regulatory changes and mandating compliance as appropriate.

- Manufacturers and operators, with oversight from the FAA, should monitor the effectiveness of corrective action and the safety management process.

Recommendation 4-1. In parallel with efforts to make appropriate regulatory changes, the FAA should expeditiously negotiate binding letters of agreement with manufacturers and operators to implement as much of the recommended safety management process as possible.

Recommendation 4-2. As the recommended safety management process is implemented, the FAA should eliminate internal efforts to collect and store data for aircraft manufactured by companies with whom agreements have been reached in accordance with Recommendation 4-1. Resources currently used for those purposes should be redirected to AIR's other safety-related functions.

Recommendation 4-3. Manufacturers should establish aviation safety database management systesms (DBMSs) using the state-of-the-art data management technologies that are best suited to continued airworthiness applications. The most suitable type of DBMS currently available is the object-relational DBMS.

Recommendation 4-4. Consistent with regulatory procedures, the FAA should develop a more accurate methodology for assessing the costs and benefits of potential airworthiness directives and other rulemaking actions. In particular, the FAA should work with industry to develop commonly accepted models for estimating time and cost.

Recommendation 4-5. To eliminate the regulatory backlog and the ambiguities about implementing airworthiness actions of foreign regulatory authorities, the FAA should expeditiously determine what regulatory action, if any, it will propose in response to foreign airworthiness actions. The FAA should initiate its regulatory response no later than two weeks after receiving notice of a foreign airworthiness action.

Major Recommendation 3. AIR should promote aircraft safety by certifying the competency of applicants' design organizations rather than relying on the FAA's ability to detect design deficiencies through spot checks. The FAA should work with industry and Congress to obtain legislative and regulatory authority in a timely fashion to do the following:

- Certificate and rate approved design organization (ADOs) and invest them with the responsibility for ensuring that applications for type certificates, type certificate amendments, supplemental type certificates, technical standard order authorizations, and parts manufacturer approvals comply with applicable airworthiness standards. ADOs would be required to have the technical capabilities necessary for competently approving designs only within the limitations of their rating.
- Require ADOs and holders of production certificates to collect and analyze relevant safety data received from operators and to define corrective action in the event unsafe conditions are detected.
- Require applicants for design approvals to either hold an ADO certificate or employ the services of an ADO.
- As an interim step, give higher priority to the ongoing rulemaking action that would increase organizational delegation to manufacturers of large aircraft and engines under the FAA's current legislative authority. The FAA already uses this authority to grant organizational delegation to manufacturers of small aircraft and engines.

CHAPTER 5

Human Factors

Finding 5-1. Maintaining situational awareness is the key to preventing the vast majority of serious incidents and accidents associated with human error.

Major Recommendation 4. The FAA should support and accelerate efforts (1) to define the minimum data required by the flight crew to maintain adequate situational awareness during all phases of flight and reasonable emergency scenarios and (2) to determine how this data can be presented most effectively.

Recommendation 5-1. The FAA should ensure that its human factors projects, especially the FAA Human Factors Study Group, include strong representation in the fields of cognitive science and basic neuroscience.

Recommendation 5-2. Advances in understanding human factors should be quickly applied to the key task of reducing the role of human errors in incidents and accidents, particularly with regard to improving the situational awareness of operational personnel and improving the effectiveness of maintenance personnel. The FAA should strongly support its Human Factors Study Group and other projects that contribute to this task.

CHAPTER 6

Barriers

Major Recommendation 5. In order for AIR to contribute as much as possible to improvements in aviation safety, the FAA—in partnership with industry, Congress, the Department of Transportation, and other involved parties—must take aggressive action to overcome barriers associated with the following:

- external pressures and influences faced by the FAA
- coordination and communications within the FAA
- legal issues
- the rulemaking process
- the economic impact of proposed changes to the safety management process

Finding 6-1. Following some highly publicized accidents, there is a technically unjustified loss of public confidence, which leads to political pressure and a counterproductive atmosphere of crisis management in the FAA that interferes with ongoing efforts by government and industry to improve aviation safety.

Recommendation 6-1. As a first step towards reducing the negative impact of external pressure on the safety management process, the FAA should work with other responsible agencies to educate the public more fully about ongoing efforts to improve aviation safety. Fully addressing this issue is likely to require major organizational changes, such as the establishment of a senior interagency communications or safety management board, that were beyond the scope of this study.

Recommendation 6-2. The FAA should develop a process to facilitate communications and improve coordination among offices within AIR and between AIR and the Flight Standards Service. For example, the Associate Administration for Regulation and Certification could establish a central coordinating office to facilitate the exchange of continued airworthiness information within the FAA and the dissemination of complete and consistent information to industry.

Recommendation 6-3. The FAA should initiate regulatory action, legislative action (through the Congress), and/or letters of understanding with manufacturers, operators, pilot organizations, and others to serve the public interest and improve safety by encouraging the voluntary sharing of safety data. This may involve limiting enforcement action based on voluntarily shared data and protecting such data from release to other parties.

Recommendation 6-4. Efforts to reform the Aviation Rulemaking Advisory Committee (ARAC) should (1) establish more timely and effective processes and (2) encourage the assignment of industry and FAA personnel who have the expertise to develop well written notices of proposed rulemaking (NPRMs) and final rules and the influence necessary for building broad support for documents approved by the ARAC.

Finding 6-2. It quite often takes 5 to 10 years to issue new regulations or modify existing regulations. This is an important safety issue because it constrains the ability of the rulemaking process to improve aviation safety. The FAA is in the process of reforming its internal procedures, including ARAC procedures. This is a positive first step, but much more needs to be done in this area.

Recommendation 6-5. The FAA should make the rulemaking process substantially more responsive by convincing the Department of Transportation, other executive branch agencies, and Congress to modify legislation, directives, and regulations to allow major changes in the current process.

Recommendation 6-6. The FAA should work with industry to develop confidence in the cost-benefit analyses used to justify changes in the safety management process. The FAA should also subsidize pilot projects by operators and manufacturers to validate the cost effectiveness of new systems for data collection, database management, and analysis.

CHAPTER 7

Small Airplanes and Rotorcraft

Major Recommendation 6. Plans to implement the recommended safety management process within the small airplane and rotorcraft communities should be developed in cooperation with small airplane and rotorcraft operators, manufacturers, and associations of operators and manufacturers. The FAA should establish cooperative agreements that define the roles of individual operators, individual manufacturers, their associations, and AIR. These agreements should define the following:

- responsibilities of operators for submitting data
- responsibilities of operators, manufacturers, associations of operators and manufacturers, and AIR for data collection, database management, risk analysis, risk management/action, and monitoring effectiveness
- processes for the routine exchange of data and risk analysis results between operators, manufacturers, associations, and AIR to facilitate effective risk management/action
- a publicity program to inform the small airplane and rotorcraft communities of the new safety management process

Recommendation 7-1. The FAA should conduct separate safety assessments for each segment of the general aviation community to identify the continued airworthiness problems of greatest significance as a function of the type of operation, class of aircraft, and experience level of the pilots.

Recommendation 7-2. AIR, in conjunction with the original equipment manufacturers of military surplus helicopters, should take timely action to define specific guidance for aircraft certification offices and industry to use in evaluating the airworthiness of military surplus helicopters in accordance with current regulatory standards.

APPENDIX B

Biographical Sketches of Committee Members

James G. O'Connor (chair) is a former president of Pratt & Whitney, which designs and builds engines for commercial, military, and general aviation aircraft. His 34-year career started in engineering and included key assignments in engineering, customer support, program management, manufacturing operations, and general management. He was involved in both military and commercial programs and businesses. His engineering assignments included development and certification of key commercial engines for Boeing and Douglas aircraft companies. In 1989, Mr. O'Connor became the chief executive for Pratt & Whitney, directing all of the aircraft engine manufacturer's $7 billion operations. He retired from Pratt & Whitney in 1993. Mr. O'Connor is currently chairman of the Board of Trustees of Embry-Riddle Aeronautical University. He is a member of the National Academy of Engineering, the National Research Council Aeronautics and Space Engineering Board, the Connecticut Academy of Science and Engineering, the President's Advisory Council at Clemson University, and the Wings Club. Mr. O'Connor earned B.S. and M.S. degrees in mechanical engineering from Clemson University and Rensselaer Polytechnic Institute, respectively. He also completed the Executive Management Development Program at Rensselaer Polytechnic Institute.

M. Craig Beard retired from the Federal Aviation Administration (FAA) in March 1996, after receiving the FAA's Distinguished Service Award for 33 years of government and military service. After joining the FAA as an aeronautical engineer, Mr. Beard held positions in the aircraft certification regulatory program at progressive levels of responsibility in Fort Worth, Texas; Brussels, Belgium; Los Angeles, California; and Washington, D.C. His service included 14 years as director of the Aircraft Certification Service and the preceding Office of Airworthiness. Just prior to leaving the FAA, Mr. Beard served nearly three years as director of the FAA's Asia Pacific Office, headquartered in Singapore. Throughout his FAA career, Mr. Beard worked extensively with the aviation safety authorities of many countries, including the European Joint Aviation Authorities and the International Civil Aviation Organization (ICAO); civil aircraft and aeronautical products manufacturers; and aviation associations in Western and Eastern Europe, the Middle East, South America, and Asia to promote aviation safety. Before joining the FAA, Mr. Beard worked as a designer and aeronautical engineer in industry and as a private consultant engineer for 10 years. He graduated from the University of Wichita (now Wichita State University), Wichita, Kansas, with a B.S. degree in aeronautical engineering in 1962; he has been a professional engineer, registered in the State of Texas, since 1965. He is a fellow of the Royal Aeronautical Society and served on the Society of Automotive Engineers (SAE) Aerospace Council. Mr. Beard has received the SAE's Franklin W. Kolk Air Transportation Progress Award (1990), the Flight Safety Foundation/Aviation Week and Space Technology Distinguished Service Award (1992), and was recognized by Aviation Week and Space Technology in its 1993 "Laurels." Mr. Beard is president of the International Federation of Airworthiness, a not-for-profit organization chartered in the United Kingdom and dedicated to improving aviation safety in all aspects of airworthiness, particularly continued airworthiness.

Eugene E. Covert was appointed the T. Wilson Professor Emeritus, Department of Aeronautics and Astronautics, Massachusetts Institute of Technology (MIT), following his retirement in 1996. In addition to his academic activities, Professor Covert was associate director of the MIT Aerophysics Laboratory until he became the director of the Gas Turbine Laboratory and department head from 1985 to 1990. His research, which was conducted through the Center of Aerodynamic Studies, has been focused on problems in unsteady fluid mechanics. Professor Covert has been both a member and chair of the U.S. Air Force Scientific Advisory Board, a member of the Aeronautics and Space Engineering Board of the National Research Council, and the 1997

Wright Brothers Lecturer. He is a member of the New York Academy of Science, the National Academy of Engineering, a fellow of the Royal Aeronautical Society, and an honorary fellow of the American Institute of Aeronautics and Astronautics. He holds B.S. and M.S. degrees in aeronautical engineering from the University of Minnesota and an Sc.D. from MIT.

Theodore E. Dumont retired after 30 years with Sikorsky Aircraft, where he was the liaison with all civil airworthiness authorities, particularly in the area of aircraft certification, and participated in the certification of the Sikorsky models S-58, S-62, S-61, S-58T, S-64, and S-76 helicopters. Prior to that time, he served for seven years as a design evaluation engineer for the U.S. Civil Aeronautics Administration (which later became the FAA), specializing in rotorcraft structures, rotor drive systems, and power-plant installations. He also served in the U.S. Air Force as the officer in charge of the Air Force facilities at Sikorsky Aircraft and Bridgeport Lycoming, which oversaw production and development contracts. Mr. Dumont has a B.S. degree in aeronautical engineering from the University of Cincinnati and received his private pilot certificate in 1940. He is currently a special advisor to the Board of Directors of the Helicopter Association International. He has received awards from both the FAA and the Helicopter Association International for his distinguished service in the cause of rotorcraft safety and for his contributions to U.S. leadership in the helicopter field.

Frank C. Fickeisen was employed by The Boeing Company for more than 40 years. He was associated with the design, test, analysis, and certification of primary and automatic flight controls on Boeing 707, 727, 747, and 767 airplanes. From 1982 to 1993, he coordinated technical work, analyses, and certification programs that led to the certification of Boeing twin-engine airplanes for extended range operations (ETOPS). He was also the Boeing technical focal point for harmonization of ETOPS regulations and many airplane system regulations between the United States and Europe. Mr. Fickeisen was a technical fellow of The Boeing Company and is currently consultant to Boeing, the Soloy Corporation, the Flight Safety Foundation, and the International Federation of Airworthiness.

Clyde Kizer is president and chief operating officer of Airbus Service Company (ASCO), the customer support subsidiary of Airbus Industrie of North America, a position he has held since December 1992. Mr. Kizer joined ASCO in January 1992 as senior vice president for product support. The Airbus Training Center in Miami, the Airbus Spares Center in Ashburn, Virginia, and customer staff support staff report to Mr. Kizer. Prior to joining Airbus, Mr. Kizer was senior vice president for airline and flight operations with Midway Airlines. From 1988 through 1990, he was vice president for engineering and maintenance with the Air Transport Association of America. He was employed by United Airlines for more than 14 years, serving as flight test captain, director of engineering, and vice president for technical services. Mr. Kizer retired from the U.S. Navy as a captain in 1982. During his naval career, he accrued more than 8,000 flight hours in combat and noncombat service. He is a graduate of the U.S. Naval Test Pilot School and a member of the Society of Experimental Test Pilots. He is a member of the SAE and the American Institute of Aeronautics and Astronautics. He has a B.S. degree in chemistry from Eastern Michigan University and has completed the Executive Management Program at Stanford University.

Dean J. Lennard retired after a 39-year career with the General Electric Company, 38 of which were in the aircraft engine business. His experience includes engineering management from initial development, certification, product engineering in support of customers' field operations, to overall project management for engines for the C5, B1, B2, F16, and ATF military aircraft and engines and nacelle systems for the B747/B767, DC10/MD11, and A300/A310/A330 commercial aircraft. During his last eight years with General Electric, Mr. Lennard was general manager of the CF6 engine family product lines, which power the commercial aircraft listed above. He holds a B.M.E. degree from the Georgia Institute of Technology and completed the General Electric Executive Development Course. He is a registered professional engineer, holds three patents, has authored various technical papers, and is a retired member of the American Institute of Aeronautics and Astronautics. Mr. Lennard was recently inducted into the GE Aircraft Engines Propulsion Hall of Fame.

Steven R. Lund is the senior principal staff engineer of flight safety investigations in flight operations for the Douglas Products Division of the Boeing Commercial Airplane Group in Long Beach, California. He has spent 34 years in the U.S. aerospace industry, the last 29 of which have been at the Douglas Aircraft Company (now Boeing). His entire career at Douglas has been devoted to flight test, flight safety, and commercial jet transport incident and accident investigation. He has been involved in the investigation/analysis of more than 180 jet transport airline accidents and more than 5,000 incidents. Mr. Lund has authored several papers on aviation safety and has prepared and taught numerous courses on airline accident investigation and flight safety. He is a past president of the International Society of Air Safety Investigators, Los Angeles Regional Chapter, and past chair of the International Coordinating Council of Aerospace Industries Associations for airline accident investigations. Mr. Lund has a B.S. in aeronautical engineering with a minor in advanced engineering mathematics from California State University, San Luis Obispo.

C. Julian May is president and chief operating officer of Tech/Ops International. He retired after 37 years with Delta

Air Lines, where his positions included senior vice president for technical operations, vice president for engineering, and vice president for technical operations planning and development. He has had responsibility for all technical activities at Delta, including engineering, new aircraft evaluation, quality control, maintenance, material services, facilities, and communications. He has been active in the Air Transport Association of America (ATA) as a member of the ATA Airworthiness Assurance Steering Committee, as the chairman of the Engineering, Maintenance and Materiel Council for 1989 and 1993, and as a member of the ATA Functions Review Team. Mr. May is a member of the American Institute of Aeronautics and Astronautics and a fellow of the SAE. He was appointed to the SAE Aerospace Council in 1980 and served as chairman for five years. He was a member of the SAE Board of Directors (1986 to 1989), is currently chairman of the newly formed SAE Engineering Leadership Award Committee, and is a member of the SAE Aerospace Program Office. Mr. May is on the Board of Directors of Cobb Family Resources, an honorary director for the Georgia Engineering Foundation, and a past member of the Aeronautics and Space Engineering Board. Mr. May was the first recipient of the SAE's Marvin Whitlock Award, a recipient of the ATA's Nuts and Bolts Award, and was awarded the 1994 SAE Colwell Medal. In 1992, he presented the William A. Littlewood Lecture at SAE Aerotech. Mr. May graduated from the Virginia Military Institute with a B.S. in engineering and received an M.B.A. in finance from Georgia State University.

William H. Schultz has been a professional engineer for 36 years and is currently the vice president for engineering and maintenance of the General Aviation Manufacturers Association in Washington, D.C. He was the 1993–1994 chairman of the FAA's Aviation Rulemaking Advisory Committee (ARAC). In 1994, he served as the industry representative on the FAA's team for the redesign of the Aircraft Certification Service, the product of which was incorporated into the 1996 report, "Challenge 2000 Recommendations for Future Aviation Safety Regulation." In 1995, he was the first chairman of the U.S. Industry Coalition for Harmonization and is still an active member of this committee. He is also the chairman of the ARAC Working Group for Certification Procedures for Products and Parts (FAR 21). Mr. Schultz has led several design review teams for the revalidation or redesign of products to resolve safety issues. He holds an M.S. in aeronautical engineering from Wichita State University, an M.S. in mechanical engineering from Michigan State University, and has completed the Raytheon Advanced Management Training Course.

Nozer D. Singpurwalla, professor of operations research, professor of statistics, and distinguished research professor of the Department of Operations Research at the George Washington University, is the director of the George Washington University's Institute for Reliability and Risk Analysis. Professor Singpurwalla is a fellow of the American Statistical Association, the Institute of Mathematical Statistics, and the American Association for the Advancement of Science. He has been a visiting professor at Stanford University; Carnegie Mellon University; the University of California, Berkeley; and the Virginia Polytechnic Institute and State University. He has authored or coauthored more than 150 publications (see http://www.seas.gwu.edu/seas/institutes/irra/) on reliability theory, statistical inference, quality control, and risk analysis. For his contributions to the scientific literature, he was awarded the Wilk's Award by the U.S. Army Research Office, a senior fellowship at the National Institute for Standards and Technology, and the Rockefeller Foundation's Scholar in Residence Fellowship.

Colin Torkington is currently an air navigation commissioner for ICAO (International Civil Aviation Organization) and the alternate member for Australia on the ICAO Council, which is based in Montreal. He started his career in 1952 with Vickers-Armstrongs, Ltd., in the United Kingdom, working on Viscount, Valiant, and TSR-2 aircraft. His final position was as a senior stressman in the design office. He obtained an M.S. degree in aeronautical engineering from Cranfield University and held a private pilot's license and glider qualification. In Australia, he joined the Department of Civil Aviation as an airworthiness engineer specializing in aircraft structures. He worked on several major accident investigations and during his career undertook 54 overseas assessments and certification visits covering authorities, manufacturers, and operators in 26 countries. He has been head of Airworthiness and Operations in the Australian Civil Aviation Authority in Canberra and chairman of the ICAO Continuing Airworthiness Panel. He is a fellow of the Royal Aeronautical Society.

William Hoover, Aeronautics and Space Engineering Board liaison to the Committee on Aircraft Certification Safety Management, is the former executive vice president of the ATA and a retired major general from the U.S. Air Force. At the ATA he was responsible for all aspects of the association's activities, including development and implementation of wide-ranging airline policies. While on active duty, Maj Gen Hoover spent four years in the U.S. Air Force space program, was a combat air wing commander in Vietnam, and later served five years as deputy assistant secretary for military applications in the Department of Energy. He holds a B.S. in engineering from the U.S. Naval Academy and an M.S. in aeronautical engineering from the Air Force Institute of Technology.

APPENDIX
C

Participants in Committee Meetings

The Committee on Aircraft Certification Safety Management met five times between February and December 1997. There were also many smaller meetings attended by one or more committee members and representatives of public and private organizations involved in the aeronautics industry. The small group meetings were part of the committee's information-gathering process. In addition to committee members and staff, participants in committee meetings included the following:

Mac Armstrong, Delta Air Lines
Ann Azevedo, Federal Aviation Administration
Michael Basehore, Federal Aviation Administration
Ben Beets, Federal Aviation Administration
Yves Benoist, Airbus Industrie
Glenn Bylsma, The Boeing Company
Thomas Boudreau, Federal Aviation Administration
Henri Branting, Federal Aviation Administration
Eric Bries, Federal Aviation Administration
James Busey, General Electric Aircraft Engines
Tom Butine, The Boeing Company
Robert Carlson, The Boeing Company
Ed Clark, Office of Management and Budget
Steven Corrie, Federal Aviation Administration
Roland Crandall, General Electric Aircraft Engines
R.E. Crow, Pratt & Whitney
Todd Curtis, The Boeing Company
Ross Cusimano, Federal Aviation Administration
Jim Devany, Federal Aviation Administration
Wolfgang Didszuhn, Airbus Industrie
John Drake, National Transportation Safety Board
Ed Dunlap, Delta Air Lines
Chet Ekstrand, The Boeing Company
Chris Erikson, Erikson Air-Crane
Jerome Frechette, National Transportation Safety Board
Curtis Graeber, The Boeing Company
Buddy Guess, Delta Air Lines
H. Keith Hagy, Air Line Pilots Association, International
James Hallock, U.S. Department of Transportation
Joe Hawkins, Federal Aviation Administration
Phil Hensley, AlliedSignal Engines
Fred Herzner, General Electric Aircraft Engines
Sally Hickman, The Boeing Company
Tim Hickox, The Boeing Company
Jim Houckey, National Transportation Safety Board
Charles Huber, Federal Aviation Administration
Frank Jenson, Helicopter Association International
Ed Kupcis, The Boeing Company
William Kurtz, General Electric Aircraft Engines
Glen Lantner, Federal Aviation Administration
Alan Lea, Pratt & Whitney Canada
Alain Leroy, European Joint Aviation Authorities
Jess Lewis, Federal Aviation Administration
Bernard Loeb, National Transportation Safety Board
Robert Loftus, General Electric Aircraft Engines
Tom Longridge, Federal Aviation Administration
Bill Machado, Federal Aviation Administration
Bob Macintosh, National Transportation Safety Board
Sarah MacLeod, Aeronautical Repair Station Association
Douglas Macnair, Aircraft Owners and Pilots Association
Norm Martenson, Federal Aviation Administration
Jim Maucere, Delta Air Lines
Wright McCartney, Alaska Airlines
John (Jack) McGrath, Federal Aviation Administration
Tom McSweeney, Federal Aviation Administration
William Melvin, Air Line Pilots Association, International
Nelson Miller, Federal Aviation Administration
Stewart Miller, Federal Aviation Administration
Randy Milne, The Boeing Company
Gerard Misrai, Airbus Industrie
Adam Monsoori, The Boeing Company
Alan Moodie, Delta Air Lines
Yves Morier, European Joint Aviation Authorities
Lee Nguyen, Federal Aviation Administration
Frank Paskiewicz, Federal Aviation Administration

Scott Patterson, Alaska Airlines
Darrell Pederson, Federal Aviation Administration
Greg Phillips, National Transportation Safety Board
Sebastian Pierini, The Boeing Company
Larry Plaster, Boeing Company-Mesa
Mike Prendergast, Pratt & Whitney
G. W. Prescott, The Boeing Company
Barry Rohm, Allison Engine Company
Richard Ridge, General Electric Aircraft Engines
Donald Riggin, Federal Aviation Administration
Mike Romanowski, Pratt & Whitney
Paul Russell, The Boeing Company
Tom Sandberg, Sikorsky Aircraft
Sam Sampath, Federal Aviation Administration
Ronald Schleed, National Transportation Safety Board
Dave Shepherd, American Eurocopter
Steve Slotte, Federal Aviation Administration
Thaddee Sulocki, European Joint Aviation Authorities
John Swihart, FAA Aviation Rulemaking Advisory Committee
Carey Terasaki, Federal Aviation Administration
Janice Thaxter Pratt & Whitney
Gary Thompson, Delta Air Lines
Jim Trimberger, Alaska Airlines
Scott Turco, Delta Air Lines
Linda Walker, Federal Aviation Administration
Susan Walsh, Pratt & Whitney
Dale Warren, Aircraft Safety Subcommittee of the FAA Research Engineering and Development Advisory Committee
George Warren, Columbia Helicopters
Al Weaver, Pratt & Whitney
Ronald Wojnar, Federal Aviation Administration
Brian Yanez, Federal Aviation Administration
Chet Yantsy, Alaska Airlines

APPENDIX D

Data Requirements

The type of data collected by a safety management process should be tailored to the requirements of the risk analysis and risk management functions so enough of the right kind of data is collected but the database management system is not overwhelmed by unnecessary data. The information presented in this appendix describes a sample set of aircraft safety data. This information is not intended to limit the user's approach, which may be altered to meet individual requirements.

Data is collected from aircraft accident or incident reports, service difficulty reports, and other reports of deficiencies in the users' sphere of interest. Relevant information includes data identifiers, descriptors, symptoms, events, causes, consequences, injury/death factors, miscellaneous data, and corrective action.

IDENTIFIERS

Identifiers should include an identification number, time of occurrence, classification, location, aircraft, mission type, operator, meteorological conditions, and phase of operation.

The identification number could be a nine-digit number denoting the date of the accident or incident, followed by the level of damage, and a sequence number indicating the number of reported accidents and incidents on that day. For example, the number could be formatted as **YYMMDDLSQ**:

- **YY** = the year, **MM** = the month, **DD** = the day
- **L** = the level of damage, with **1** = destroyed/fatal, **2** = destroyed/nonfatal, **3** = substantial/fatal, **4** = substantial/nonfatal, **5** = incident, **6** = service difficulty, and **7** = other
- **SQ** = the two-digit number of the occurrences on that date

The time should be recorded in universal time and local time (24-hour clock) along with ambient lighting conditions: dawn, daylight, dusk, or night.

Classification options would be accident, incident, or service difficulty.

The location of the aircraft at the time of the problem should be recorded using the four-digit codes for the departure and arrival airports as designated in the Official Airline Guide. The latitude and longitude or the distance and direction to or from the nearest city should also be recorded.

The aircraft type, model, registration number, fuselage number, fuselage serial number, engine make and model, engine serial numbers, and flight number should be recorded.

Possible mission types include scheduled passenger flight, nonscheduled passenger flight, cargo flight, ferry flight, training flight, in-flight refueling, and flight test.

The aircraft owner, operator, lessor, airline name, and nation of registry should be recorded.

The meteorological conditions should be recorded, indicating whether visual or instrument meteorological conditions (VMC or IMC) were present.

The phase of operation should be described using the following choices: boarding, cargo loading, engine start, taxi, takeoff (including roll, rotation, initial climb), climb to cruise, cruise, in-flight refueling (including precontact, contact, and disconnect), descent, approach, landing (including flare and touchdown, roll, and touch-and-go), go-around, and deplaning. For ground operations, the phase could be parked, refueling, inspection, or towing. If the problem was related to maintenance, the phase could be overhaul, servicing, routine, or inspection.

DESCRIPTORS

In addition to identifiers, the following list of descriptors could be used to categorize similar accidents, incidents, and service difficulties in more detail for subsequent sorting:

- pre-takeoff or post-landing
- collision
- lift problem

- smoke, fire, and/or fumes
- landing and taxi
- landing gear and/or brakes
- thrust problems
- flight control problems
- weather
- other aircraft factors
- other personnel factors
- facilities

As many subcategories as necessary should be used to describe accurately each accident, incident, or service difficulty.

Pre-takeoff and post-landing problems can involve the ground crew, passengers, collisions with vehicles, emergency evacuations, returns to the gate, rejected takeoffs and other takeoff problems, tires, engines, bird strikes, running off the end or side of the runway, weather, darkness, using the wrong runway, overrotation and/or tail strike, underrotation, slat disagreement warnings, wet runways, loss of directional control, premature retractions of the landing gear, and spoiler position warnings.

Collisions may involve other aircraft on the ground or in flight; terrain; water; objects on the ground or water; foreign objects or domestic objects on the ground or in flight; birds impacting engines, windscreens, or structure; airport vehicles or ground equipment; and formation flying or refueling operations. For the purposes of safety management and accident prevention, near misses would be treated the same way as collisions.

Lift problems may be caused by the loss of one or more lifting surfaces; contaminated lifting surfaces; improper configurations; or asymmetric, partial, or erroneous deployment of slats, flaps, or spoilers.

Accidents, incidents, and service difficulties involving smoke, fire, and/or fumes should be classified by the source of the problem (electrical systems, flammable fluids, cargo, auxiliary power unit, or other system) and the location of the problem within the aircraft (cockpit, cabin, galley, lavatory, cargo compartment).

Landing and taxi accidents may be classified as unscheduled landings, gear-up landings, excursions off the end or side of the runway, brake malfunctions or difficulties, go-arounds, short landings, long landings, overweight landings, hard landings, problems from contaminated runways, using the wrong runway, ditching, loss of directional control, or runway contact by nacelle, wing, tail, or fuselage.

Landing gear and brake problems include collapse of the landing gear, tire failure, wheel failure, brake failure, asymmetric braking, main landing gear up or unlocked, nose landing gear up or unlocked, false gear indication, antiskid system failure, steering failure, strut failure, emergency extension (gear free-fall), asymmetric gear extension, and tire damage by foreign objects.

Thrust problems include in-flight engine shutdowns, failure or asymmetric deployment of thrust reversers, inadvertent thrust reverser deployment in flight or on the ground, engine flameout, engine failure that is not contained by the engine case and/or the nacelle, engine fire warning, engine separation, high exhaust gas temperature, engine stab or surge, engine power loss, multiple engine failure, foreign object damage to engine, engine overspeed, and abnormalities involving the oil system, throttles, gear box, or fuel.

Flight control problems include gross weight and center-of-gravity problems, jammed or locked controls, aircraft stall, instrument error or false indications, wake turbulence, buffet, or vibrations caused by structural failures, improper actions by the pilot or autopilot, uncommanded actuation of control surfaces, adverse weather, or other system cause.

Weather-induced problems include ice formation, ice shedding, turbulence, lightning strike and/or static discharge, clouds, winds (tailwinds, headwinds, and crosswinds), thunderstorms, wind shear, microbursts, fog, haze, and all forms of precipitation: rain, heavy rain, freezing rain, snow, slush, hail, etc.

Other aircraft factors include depressurization, emergency descent, fuselage shell opening, warning indications, uncommanded actuation of aircraft systems or controls, oxygen system problems, hazardous cargo, rotating machinery failure, multiple failures, air conditioning and pressurization problems, pneumatic system malfunctions, hydraulic system malfunctions, electrical system malfunctions, fuel system problems, exceeding "g" limits, separation of parts in flight, fluid seepage and spills, blue ice, jettisoning of fuel, leakage of fluids from cargo, loose cargo, improper activation of fire extinguishers, malfunctions of items on the minimum equipment list.

Other personnel factors include problems caused by the flight crew, cabin crew, passengers, ground crew, air traffic controllers, maintenance crews, and others because of confusion, fatigue, inadequate coordination, poor communications, terrorist acts, etc.

Facility problems include airfield obstacles, inadequate braking because of runway contamination or other reasons, poor lighting, inadequate signage, problems with landing or navigational aids, malfunctions of air traffic control equipment, inadequate emergency crash/fire/rescue equipment, etc.

SYMPTOMS

At some point, almost all accidents, incidents, and service difficulties generate symptoms that indicate something has gone wrong.

Visual symptoms include instrument indications, warning or advisory lights, and observations of smoke, fire, or other abnormal conditions.

Aural symptoms include horns, bells, and verbal warnings generated by the central aural warning system; warnings and alarms generated by other systems, such as the ground proximity warning system or fire and smoke alarms; and abnormal sounds generated by malfunctioning equipment.

Tactile symptoms include aircraft "g" loading, stick shaker, control forces, heat, cold, pressure change, electrical shock, and vibration.

Olfactory symptoms include smoke or fumes from electrical systems, oil, kerosene, food, rubber, or other materials.

LINKS IN THE CHAIN OF EVENTS

Accidents and, to a lesser extent, incidents and service difficulties typically involve a chain of events. Each event or link in the chain should be categorized and described.

Personnel errors should be recorded by identifying the individuals involved (captain, first officer, second officer, cabin crew, ground crew, maintenance crew, passenger, other flight crew, etc.), describing the errors, such as failure of the flight crew to properly secure passengers, take immediate action, follow air traffic control instructions, use checklists, compensate for wind, maintain direction control, go around, monitor weather, monitor instruments, recover from unusual attitude, inform cabin crew or other flight crew, see and avoid other aircraft, accurately estimate altitude, interpret instrument readings, or maneuver the aircraft in accordance with approved procedures. Personnel errors include voluntary acts that are poorly performed or failures to act when action is appropriate.

Problems with aircraft, engines, and aircraft systems should be identified using the standard Air Transport Association chapter number. If possible, the make, model, name, and serial number should be recorded, along with failure mode, total service time, and total time since overhaul, as appropriate.

Events related to the air traffic management system include air traffic control functions, such as clearance delivery, ground control guidance, tower operations, approach and departure control, guidance from air route traffic control centers; air traffic control directives, facilities, navigation aids; and weather reporting.

Environmental conditions of interest include poor visibility because of darkness, the position of the sun, or whiteout; clear-air turbulence; birds; volcanic ash; dust; and weather.

CAUSE FACTORS

Each link in the chain of events has one or more cause factors that describe the underlying reasons, problems, deficiencies, or acts that caused the event.

Personnel errors and human factors problems may be caused by inadequate preparation or supervision, poor judgment, poor execution, inadequate crew rest, carelessness, improper use of equipment, alcohol or other drugs, exceeding aircraft operating limitations, improper maintenance, improper aircraft modifications, failure to use proper safety precautions, slow response when immediate action is needed, inadequate procedures, and failure to execute proper procedures.

Cause factors associated with aircraft, engines, and systems include abnormalities in the design, manufacture, maintenance, operation, or operating environment of the aircraft or its systems, including inadequate overhaul or inspection, foreign object damage, or the effects of failures of other aircraft systems.

Cause factors associated with air traffic management include deficient regulations, directives, or procedures; problems with air traffic control and runway facilities, such as navigational aids, communications systems, crash/fire/rescue equipment, runways, runway lighting, and taxiways; and personnel errors associated with the above.

Environmental cause factors include inaccurate forecasts of hazardous environmental conditions, failures to report hazardous conditions, or inappropriate responses to actual or forecast hazardous conditions.

Maintenance-related cause factors include improperly performed maintenance and inadequate maintenance procedures and plans.

Combinations of factors and cascading cause-and-effect sequences are also important and should be carefully examined and recorded.

CONSEQUENCES

Accidents, incidents, and service difficulties can lead to abnormal actions by the air crew, aircraft damage, and personnel injuries.

Actions by the flight crew include air turn back, rejected takeoff, return to gate, emergency descent, unscheduled landing, dumping fuel, in-flight shutdown of one or more engines, delayed departure, emergency evacuation, and evasive action.

Levels of aircraft damage include destroyed, substantial, minor, none, engine only, or property.

Personnel losses should be described in terms of the numbers of passengers, cabin crew, flight crew, ground crew, and other personnel injured and the extent of injuries: fatal, serious, minor, or none.

If the aircraft impacts ground or water, it should be noted whether the aircraft was under control at the time of impact.

For accidents, the official determination of whether it was survivable or nonsurvivable should be entered.

INJURY/DEATH FACTORS

Additional information should be recorded about factors associated with personnel injuries in case additional remedial action is needed.

Crash and non-crash-related injury factors should be recorded separately. Non-crash-related injuries may be caused by falls, crushing, sharp edges, dropped objects, turbulence, electric shocks, electric burns, open flames, toxic smoke or fumes, hot surface or liquids, radiation, or toxic chemicals. Crash-related injuries may be caused by impact forces, inadequate seat or occupant restraints, fires, crushing, unrestrained objects, post-crash fires, toxic gases, drowning after ditching, or lack of post-crash egress because of physical blockages, smoke, flame, or inoperative emergency exits.

Crash and non-crash-related death factors should be recorded and categorized in the same manner as the injury factors.

Other crash factors that should be recorded include the passenger load factor in percent, the passenger distribution in the cabin, and impact "g" force.

MISCELLANEOUS DATA

The aircraft history should be recorded in terms of the number of flight hours and landings, along with a record of the flight crew's experience and demographics. A narrative of the accident, incident, or service difficulty with analyst's comments should also be included.

CORRECTIVE ACTION

When available and as appropriate, the corrective action should be recorded. Options include airworthiness directives, alert service bulletins, service bulletins, "all operator" letters, operational occurrence reports, and/or official recommendations.

APPENDIX E

Probability and Reliability Analysis

WHAT IS PROBABILITY?

Probability is a number between 0 and 1 that expresses a degree of uncertainty about whether an event, such as an accident, will occur. A logically impossible event is assigned the number 0, and a logically certain event is assigned the number 1. The axioms of probability tell us how to combine various uncertainties.

Interpretations of Probability

There are at least four interpretations of probability:

1. classical (equally likely)
2. logical (the "necessarist" position)
3. relative frequency (objectivistic)
4. personalistic (subjectivistic)

The classical interpretation is based on the "principle of insufficient reason" and was advocated by the determinists Bernoulli, Laplace, De Moivre, and Bayes. This interpretation has limited applicability and is now subsumed under the personalistic interpretation.

The logical interpretation was favored by logicians, such as Keynes, Reichenbach, and Carnap, and is currently out of vogue.

The relative frequency interpretation is used by many statisticians and is currently the most favored. This interpretation requires the conceptualization of an infinite collective and is not applicable in one-of-a-kind situations.

The personalistic interpretation is more universal and incorporates engineering and other knowledge. This interpretation is popular in many applications, including risk analysis and safety analysis.

Axioms of Probability: Dependence and Independence

All the interpretations of probability have a common set of axioms that tell us how to combine probabilities of different events. But why should risk analysts be interested in such mathematical details? Because one of the axioms pertains to the notion of dependence (and independence), a matter that is not carefully addressed by either the FAA or industry.

Consider two events ε_1 and ε_2:

$P(\varepsilon)$ denotes the probability that an event ε will occur
$P(\varepsilon_i) = P$ (basic event "i" occurs)
$= P$ (component "i" fails during its mission), $i = 1, 2$.

For example, let

ε_1 = engine 1 fails during flight.
ε_2 = engine 2 fails during flight.

Then, the axioms are:

i) $0 \le P(\varepsilon_i) \le 1,\ i = 1, 2$ (convexity)
ii) $P(\varepsilon_1 \text{ or } \varepsilon_2) = P(\varepsilon_1) + P(\varepsilon_2) - P(\varepsilon_1 \text{ and } \varepsilon_2)$ (addition)
iii) $P(\varepsilon_1 \text{ and } \varepsilon_2) = P(\varepsilon_1 \mid \varepsilon_2) \cdot P(\varepsilon_2)$
$= P(\varepsilon_1) \cdot P(\varepsilon_2)$ if $\varepsilon_1 \perp \varepsilon_2$ (multiplication)[1]

FAULT TREE ANALYSIS

Fault tree analysis is an engineering tool that, among other things, can help assess probabilities of the occurrence of undesirable events. The undesirable event is called the "top event."

The "and" and "or" gates of a fault tree correspond to the "and" and the "or" functions in the axioms (or the calculus) of probability. At the very bottom of the tree are "basic events," which usually correspond to equipment failures. Fault trees are similar to block diagrams of a system. Examples are illustrated in Figures E-1 through E-4.

[1] $\varepsilon_1 \perp \varepsilon_2$ means ε_1 is independent of ε_2.

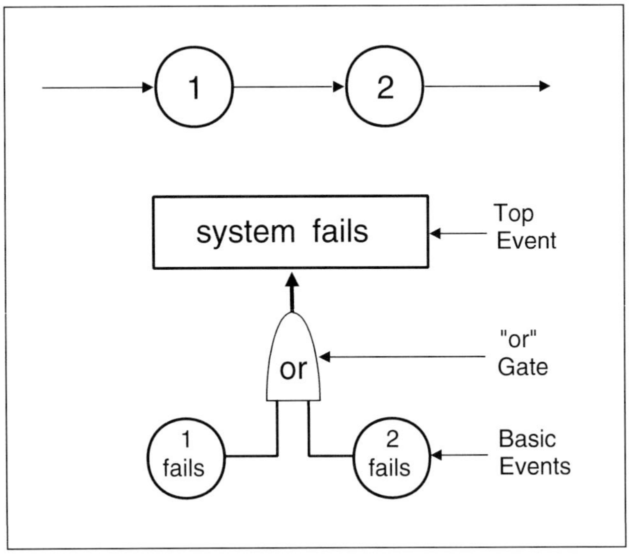

FIGURE E-1 Series system.

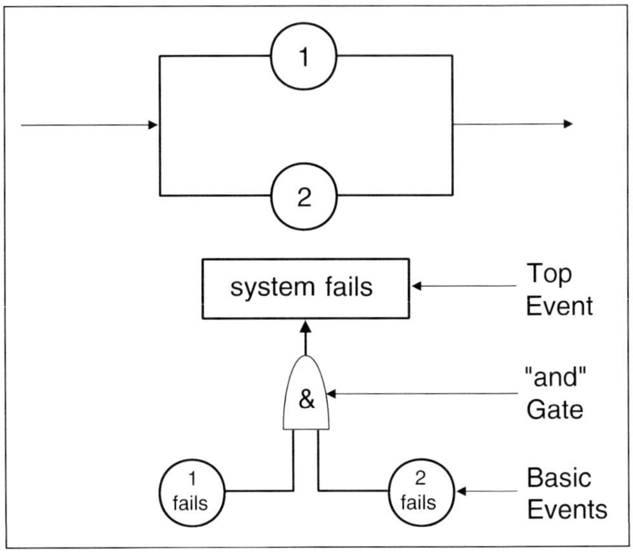

FIGURE E-2 Parallel system.

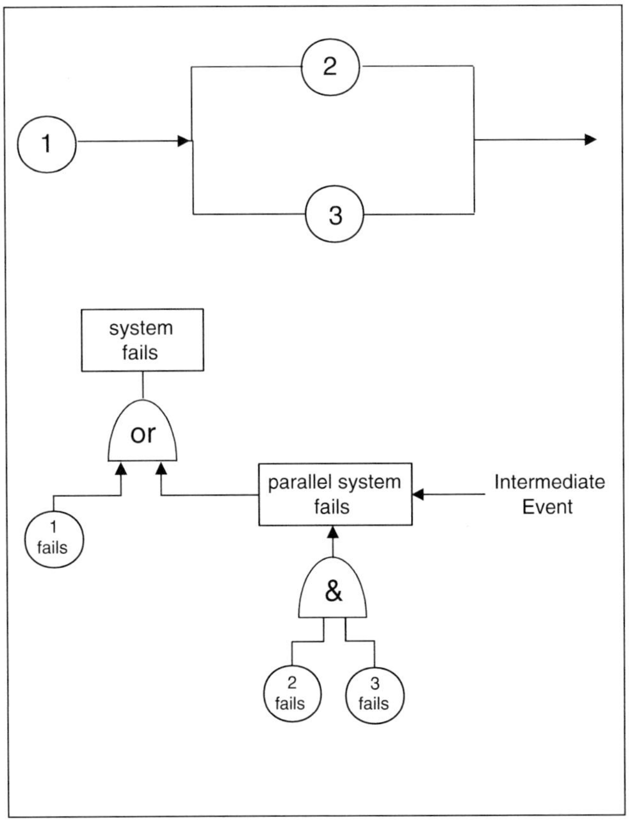

FIGURE E-3 Series-parallel system.

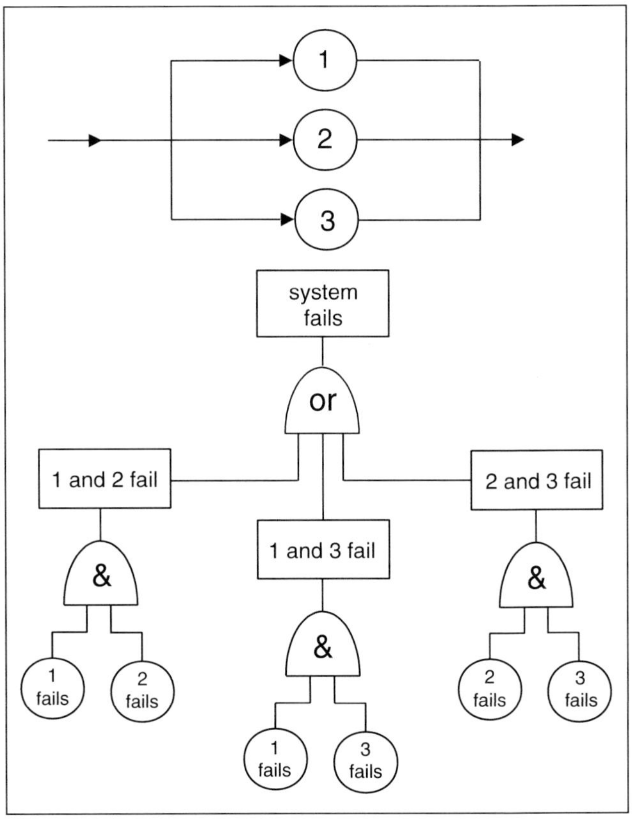

FIGURE E-4 Two-out-of-three system.

Assessing Top Event Probabilities

Let $P(T.E.) = P$ (top event occurs)
$= P$ (system fails during its mission).

How do we obtain $P(T.E.)$? This is the subject of reliability analysis wherein mathematical models, expert judgment, failure data, and maintenance come into play. Consider the following cases.

Series System with "Independence"

$$P(T.E.) = P(\varepsilon_1) + P(\varepsilon_2) - P(\varepsilon_1) \times P(\varepsilon_2)$$

When ε_1 and ε_2 are dependent, we need sophisticated reliability models to evaluate $P(T.E.)$, as discussed below.

Parallel System with "Independence"

$$P(T.E.) = P(\varepsilon_1) \times P(\varepsilon_2)$$

Series-Parallel System with Independence

$$P(T.E.) = P(\varepsilon_1) + P(\varepsilon_2) \times P(\varepsilon_3) - P(\varepsilon_1) \times P(\varepsilon_2) \times P(\varepsilon_3)$$

Two-out-of-Three System

$$P(T.E.) = P(\varepsilon_1) \times P(\varepsilon_2) + P(\varepsilon_1) \times P(\varepsilon_3) + P(\varepsilon_2) \times P(\varepsilon_3) - 2\,P(\varepsilon_1) \times P(\varepsilon_2) \times P(\varepsilon_3)$$

ASSUMPTIONS OF INDEPENDENCE

In general, assuming independence under an "and" gate *underestimates* the probability of the top event (an accident or incident). Conversely, assuming independence under an "or" gate *overestimates* the probability of the top event. The assumption of independence is an idealization often made routinely because it simplifies the analysis, but the consequences can be severe. Thus, to avoid a false sense of security, it is important that risk analysis procedures and documents used by both industry and the FAA treat dependence/independence properly.

EXAMPLE INCORPORATING DEPENDENT FAILURES

Consider a twin engine aircraft. To calculate the probability that both engines will fail by the time the aircraft accumulates some number of operating hours, τ, it is necessary to develop a probability model. A simple model is to assume that the time to engine failure has an exponential distribution with failure rate, λ, and that the failure rates are independent of each other. For that case, the probability that both engines will fail simultaneously is:

$$P(\varepsilon_1 \,\&\, \varepsilon_2) = \left(1 - e^{-\lambda_1\tau}\right)\left(1 - e^{-\lambda_2\tau}\right)$$
$$= \left(1 - e^{-\lambda\tau}\right)^2 \ \text{if}\ \lambda_1 = \lambda_2 = \lambda$$

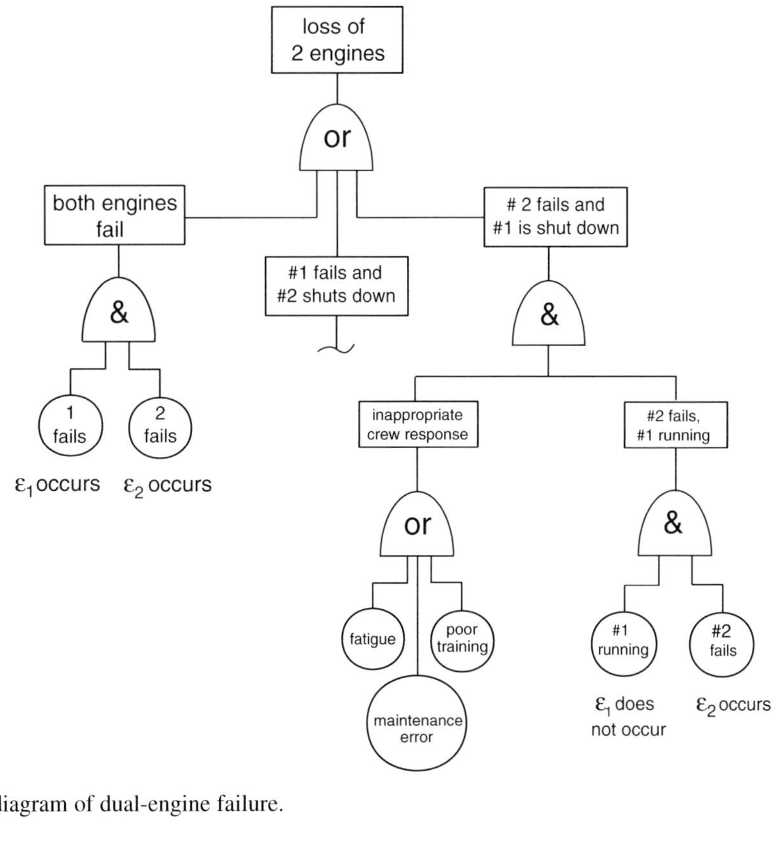

FIGURE E-5 Fault tree diagram of dual-engine failure.

A more sophisticated approach is to consider the possibility of dependent or common mode failures. For example, Figure E-5 illustrates the possibility that a failure in one engine could prompt the flight crew to shut down the functional engine, which would result in the loss of both engines even though only one engine malfunctioned. A model for common mode failures can be created via a new parameter λ^*. Now,

$$P(\varepsilon_1 \ \& \ \varepsilon_2) = 1 - 2e^{-(\lambda+\lambda^*)\tau} + e^{-(2\lambda+\lambda^*)\tau}$$

Clearly, the two probabilities are different. This shows that independence underestimates the risk of both engines failing.

APPENDIX F

Sample Legislative Amendment for Approved Design Organizations

This amendment provides a sample legislative amendment that would authorize the FAA to certificate design organizations in accordance with Major Recommendation 3. This draft is provided to facilitate quick action by the FAA and Congress.

AMENDMENTS TO 49 USC, CHAPTER 447—SAFETY REGULATION TO AUTHORIZE THE ISSUANCE OF DESIGN ORGANIZATION CERTIFICATES.

(Areas of change italicized.)

1. Amend the first sentence of §44702(a) to read:

   (a) GENERAL AUTHORITY AND APPLICATIONS.—The Administrator of the Federal Aviation Administration may issue airman certificates, *design organization certificates,* type certificates, production certificates, airworthiness certificates, air carrier operating certificates, airport operating certificates, air agency certificates, and air navigation facility certificates under this chapter.

2. Amend the title of §44704 to read:

   §44704. *Design Organization certificates,* type certificates, production certificates, and airworthiness certificates

3. Amend §44704 by redesignating paragraphs (a), (b), (c) and (d) as (b), (c), (d), and (e); and, by adding a new paragraph (a) to read:

   (a) DESIGN ORGANIZATION CERTIFICATES. The Administrator may issue a design organization certificate to authorize a design organization to certify compliance with the requirements and minimum standards prescribed under section 44701(a) for the type certification of aircraft, aircraft engines, propellers, or appliances. On receiving an application for a design organization certificate, the Administrator shall examine and rate the design organization in accordance with the regulations prescribed by the Administrator to determine that the design organization has adequate engineering, design, and testing capabilities, standards, and safeguards to ensure that the product being certificated is properly designed and manufactured, performs properly, and meets the regulations and minimum standards prescribed under section 44701(a) of this title. The Administrator shall include in a design organization certificate terms required in the interest of safety.

4. Amend redesignated §44704(b)(1) to read:

   (b) TYPE CERTIFICATES. (1) The Administrator shall issue a type certificate for an aircraft, aircraft engine, or propeller, or for an appliance specified under paragraph (2)(A) of this subsection when the Administrator finds that the aircraft, aircraft engine, propeller, or appliance is properly designed and manufactured, performs properly, and meets the regulations and minimum standards prescribed under section 44701(a) of this title. On receiving an application for a type certificate, the Administrator shall investigate the application and may conduct a hearing. The Administrator shall make or require the applicant to make, tests the Administrator considers necessary in the interest of safety. *Alternatively, the Administrator my issue a type certificate based on a certification of compliance made by a design organization certificated under section 44704(a).*

5. Amend §44709(a) to read:

   (a) REINSPECTION AND REEXAMINATION.—The Administrator of the Federal Aviation Administration may reinspect at any time a civil aircraft, aircraft engine, propeller, appliance, *design organization, production certificate holder,* air navigation facility or air agency, or airman holding a certificate issued under section 44703 of the title.

6. Amend §44711(a)(7), Prohibitions and exemptions to read:

   (a) PROHIBITIONS—A person may not—

. . .

(7) violate a term of an air agency, *design organization certificate,* or production certificate or a regulation prescribed or order issued under section 44701(a) of (b) or any of sections 44702–44716 of this title related to the holder of the certificate.

Acronyms

ACO Aircraft Certification Office
AD Airworthiness Directive
ADO Approved Design Organization
AIR Aircraft Certification Service
ARAC Aviation Rulemaking Advisory Committee

BASIS British Airways Safety Information System

CAAM Continued Airworthiness Assessment Methodologies
CFIT controlled flight into terrain

DBMS data base management system
DFDR digital flight data recorder

FAA Federal Aviation Administration
FAR Federal Aviation Regulation (as codified in Title 14 of the Code of Federal Regulations)
FOQA flight operations quality assurance

GAIN Global Analysis and Information Network

ICAO International Civil Aviation Organization

JAA Joint Aviation Authorities (an organization of 27 European civil aviation authorities)

MIDO Manufacturing Inspection District Office

NPRM notice of proposed rulemaking
NTSB National Transportation Safety Board

OMB Office of Management and Budget

PMA parts manufacturer approval

QAR quick access recorder

STC supplemental type certificate

TSOA technical standard order authorization